T0235239

Lecture Notes in Bioinformatics 10252

Subseries of Lecture Notes in Computer Science

Daniel Figueiredo · Carlos Martín-Vide
Diogo Pratas · Miguel A. Vega-Rodríguez (Eds.)

Algorithms for Computational Biology

4th International Conference, AlCoB 2017
Aveiro, Portugal, June 5–6, 2017
Proceedings

Springer

Editors
Daniel Figueiredo
University of Aveiro
Aveiro
Portugal

Carlos Martín-Vide
Rovira i Virgili University
Tarragona
Spain

Diogo Pratas
University of Aveiro
Aveiro
Portugal

Miguel A. Vega-Rodríguez
University of Extremadura
Caceres
Spain

ISSN 0302-9743
Lecture Notes in Bioinformatics
ISBN 978-3-319-58162-0
DOI 10.1007/978-3-319-58163-7

ISSN 1611-3349 (electronic)

ISBN 978-3-319-58163-7 (eBook)

Library of Congress Control Number: 2017938562

LNCS Sublibrary: SL8 – Bioinformatics

Printed on acid-free paper

This Springer imprint is published by Springer Nature
The registered company is Springer International Publishing AG
The registered company address is: Gewerbestrasse 11, 6330 Cham, Switzerland

Preface

These proceedings contain the papers that were presented at the 4th International Conference on Algorithms for Computational Biology (AlCoB 2017), held in Aveiro, Portugal, during June 5–6, 2017.

The scope of AlCoB includes topics of either theoretical or applied interest, namely:

- Exact sequence analysis
- Approximate sequence analysis
- Pairwise sequence alignment
- Multiple sequence alignment
- Sequence assembly
- Genome rearrangement
- Regulatory motif finding
- Phylogeny reconstruction
- Phylogeny comparison
- Structure prediction
- Compressive genomics
- Proteomics: molecular pathways, interaction networks
- Transcriptomics: splicing variants, isoform inference and quantification, differential analysis
- Next-generation sequencing: population genomics, metagenomics, metatranscriptomics
- Microbiome analysis
- Systems biology

AlCoB 2017 received 24 submissions. Most papers were reviewed by three Program Committee members. There were also a few external reviewers consulted. After a thorough and vivid discussion phase, the committee decided to accept ten papers (which represents an acceptance rate of about 42%). The conference program included three invited talks and some poster presentations of work in progress.

The excellent facilities provided by the EasyChair conference management system allowed us to deal with the submissions successfully and to handle the preparation of these proceedings in time.

We would like to thank all invited speakers and authors for their contributions, the Program Committee and the external reviewers for their cooperation, and Springer for its very professional publishing work.

March 2017

Daniel Figueiredo
Carlos Martín-Vide
Diogo Pratas
Miguel A. Vega-Rodríguez

Organization

AlCoB 2017 was organized by the Center for Research and Development in Mathematics and Applications, CIDMA, University of Aveiro, Portugal, the Institute of Electronics and Informatics Engineering of Aveiro, IEETA, University of Aveiro, Portugal, and the Research Group on Mathematical Linguistics, GRLMC, Rovira i Virgili University, Tarragona, Spain.

Program Committee

Can Alkan	Bilkent University, Turkey
Stephen Altschul	National Institutes of Health, USA
Yurii Aulchenko	PolyOmica, Netherlands
Timothy L. Bailey	University of Nevada, USA
Bonnie Berger	Massachusetts Institute of Technology, USA
Philipp Bucher	Swiss Federal Institute of Technology, Switzerland
Ken Chen	University of Texas MD Anderson Cancer Center, USA
Julio Collado-Vides	National Autonomous University of Mexico, Mexico
Eytan Domany	Weizmann Institute of Science, Israel
Dmitrij Frishman	Technical University of Munich, Germany
Terry Furey	University of North Carolina, USA
Olivier Gascuel	Pasteur Institute, France
Debashis Ghosh	University of Colorado, USA
Susumu Goto	Kyoto University, Japan
Osamu Gotoh	Institute of Advanced Industrial Science and Technology, Japan
Artemis Hatzigeorgiou	University of Thessaly, Greece
Javier Herrero	University College London, UK
Karsten Hokamp	Trinity College Dublin, Ireland
Fereydoun Hormozdiari	University of California, USA
Kazutaka Katoh	Osaka University, Japan
Lukasz Kurgan	Virginia Commonwealth University, USA
Gerton Lunter	University of Oxford, UK
Carlos Martín-Vide	Rovira i Virgili University, Spain, Chair
Zemin Ning	Wellcome Trust Sanger Institute, UK
William Stafford Noble	University of Washington, USA
Cedric Notredame	Center for Genomic Regulation, Spain
Christos Ouzounis	Centre for Research and Technology Hellas, Greece
Manuel C. Peitsch	Philip Morris International, Switzerland
Matteo Pellegrini	University of California, USA
Graziano Pesole	University of Bari, Italy
David Posada	University of Vigo, Spain

Knut Reinert	Free University of Berlin, Germany
Peter Robinson	The Jackson Laboratory, USA
Julio Rozas	University of Barcelona, Spain
David Sankoff	University of Ottawa, Canada
Alejandro Schäffer	National Institutes of Health, USA
Xinghua Shi	University of North Carolina, USA
Nicholas D. Socci	Memorial Sloan Kettering Cancer Center, USA
Alexandros Stamatakis	Heidelberg Institute for Theoretical Studies, Germany
Granger Sutton	J. Craig Venter Institute, USA
Kristel Van Steen	University of Liège, Belgium
Arndt Von Haeseler	Center for Integrative Bioinformatics Vienna, Austria
Kai Wang	Columbia University, USA
Ioannis Xenarios	Swiss Institute of Bioinformatics, Switzerland
Jinn-Moon Yang	National Chiao Tung University, Taiwan
Shibu Yooseph	University of Central Florida, USA
Mohammed J. Zaki	Rensselaer Polytechnic Institute, USA
Daniel Zerbino	European Bioinformatics Institute, UK
Weixiong Zhang	Washington University in St. Louis, USA
Zhongming Zhao	University of Texas Health Science Center at Houston, USA

Additional Reviewers

Xian Fan
Nam S. Vo
Chunfang Zheng

Organizing Committee

Diana Costa	CIDMA, Aveiro
Daniel Figueiredo	CIDMA, Aveiro (Co-chair)
Carlos Martín-Vide	Tarragona (Co-chair)
Manuel A. Martins	CIDMA, Aveiro
Manuel Jesús Parra Royón	Granada
Armando J. Pinho	IEETA, Aveiro
Diogo Pratas	IEETA, Aveiro (Co-chair)
David Silva	London
Miguel A. Vega-Rodríguez	Cáceres

Contents

Sequence Analysis and Other Biological Processes

Invited Talks

Biomedical Applications of Prototype Based Classifiers and Relevance Learning

Michael Biehl[✉]

Johann Bernoulli Institute for Mathematics and Computer Science,
University of Groningen, P.O. Box 407, 9700 AK Groningen, The Netherlands
m.biehl@rug.nl

Abstract. In this contribution, prototype-based systems and relevance learning are presented and discussed in the context of biomedical data analysis. Learning Vector Quantization and Matrix Relevance Learning serve as the main examples. After introducing basic concepts and related approaches, example applications of Generalized Matrix Relevance Learning are reviewed, including the classification of adrenal tumors based on steroid metabolomics data, the analysis of cytokine expression in the context of Rheumatoid Arthritis, and the prediction of recurrence risk in renal tumors based on gene expression.

Keywords: Prototype-based classification · Learning Vector Quantization · Relevance learning · Biomedical data analysis

1 Introduction

The development of novel technologies for biomedical research and clinical practice have led to an impressive increase of the amount and complexity of electronically available data. Large amounts of potentially high-dimensional data are available from different imaging platforms, genomics, proteomics and other *omics* techniques, or longitudinal studies of large patient cohorts. At the same time there is a clear trend towards personalized medicine in complex diseases such as cancer or heart disorders.

As a consequence, an ever-increasing need for powerful automated data analysis is observed. Machine Learning can provide efficient tools for tasks including problems of unsupervised learning, e.g. in the context of clustering, and supervised learning for classification and diagnosis, regression, risk assessment or outcome prediction.

In biomedical and more general life science applications, it is particularly important that algorithms provide *white box* solutions. For instance, the criteria which determine the outcome of a particular diagnosis system or recommendation scheme, should be transparent to the user. On the one hand, this increases the acceptance of automated systems among practitioners. In basic research, on the other hand, interpretable systems may provide novel insights into the nature of the problem at hand.

D. Figueiredo et al. (Eds.): AlCoB 2017, LNBI 10252, pp. 3–23, 2017.
DOI: 10.1007/978-3-319-58163-7_1

Prototype-based classifiers constitute a powerful family of tools for supervised data analysis. These systems are parameterized in terms of class-specific representatives in the original feature space and, therefore, facilitate direct interpretation of the classifiers. In addition, prototype-based systems can be further enhanced by the data-driven optimization of adaptive distance measures. The framework of relevance learning increases the flexibility of the approaches significantly and can provide important insights into the role of the considered features.

In Sect. 2, the basic concepts of prototype based classification is introduced with emphasis on the framework of Learning Vector Quantization (LVQ). The use of standard and unconventional distances is briefly discussed before relevance learning is introduced in Sect. 2.5. Emphasis is on the so-called Generalized Relevance Matrix LVQ (GMLVQ). Section 3 presents the application of GMLVQ in several relevant biomedical problems, before a brief summary is given in Sect. 4.

2   Distance-based Classification and Prototypes

Here, a brief review of distance based systems is provided. First, the concepts of Nearest Prototype Classifiers and Learning Vector Quantization (LVQ) are presented in Sects. 2.1 and 2.2. The presentation focusses on their relation to the classical Nearest Neighbor classifier. In Sect. 2.3 examples of non-standard distance measures are briefly discussed. Eventually, adaptive dissimilarities in the framework of relevance learning are introduced in Sect. 2.4.

2.1   Nearest Prototype Classifiers

Similarity based schemes constitute an important and successful framework for the supervised training of classifiers in machine learning [10,14,31,55]. The basic idea of comparing observations with a set of reference data is at the core of the classical Nearest-Neighbor (NN) or, more generally, k-Nearest-Neighbor (kNN) scheme [14,31,55,66]. This very popular approach is easy to implement and serves as an important baseline for the evaluation of alternative algorithms.

A given set of P feature vectors and associated class labels

$$I\!D = \{\mathbf{x}^\mu, y^\mu = y(\mathbf{x}^\mu)\}_{\mu=1}^{P} \quad \text{where} \quad \mathbf{x}^\mu \in I\!R^N \text{ and } y^\mu \in \{1, 2, \dots C\} \quad (1)$$

is stored as a reference set. An arbitrary feature vector or *query* $\mathbf{x} \in I\!R^N$ is then classified according to its similarity to the reference samples: The vector $\mathbf{x}$ is assigned to the class of its Nearest Neighbor in $I\!D$. Very frequently, the (squared) Euclidean distance with $d(\mathbf{x}, \mathbf{x}^\mu) = (\mathbf{x} - \mathbf{x}^\mu)^2$ is employed for the comparison. The more general kNN classifier determines the majority class membership among the k closest samples. Figure 1(a) illustrates the concept in terms of the NN-classifier.

While kNN classification is very intuitive and does not require an explicit *training phase*, an essential drawback is obvious: For large data sets $I\!D$, storage needs are significant and, moreover, computing and sorting all distances

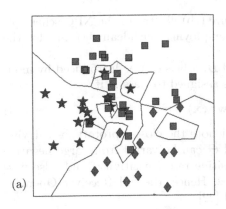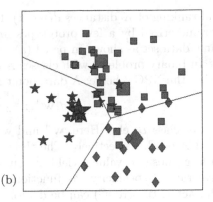

(a)   (b)

Fig. 1. Illustration of Nearest-Neighbor classification (panel a) and Nearest-Prototype classification in LVQ (panel b). The same two-dimensional data set with three different classes (marked by squares, diamonds and pentagrams) is shown in both panels. Piecewise linear decision boundaries, based on Euclidean distance are shown for the NN classifier in (a), while panel (b) corresponds to an NPC with prototypes marked by large symbols.

$d(\mathbf{x}, \mathbf{x}^{\mu})$ becomes costly, even if sophisticated bookkeeping and sorting strategies are employed. Most importantly, NN or kNN classifiers tend to realize very complex decision boundaries which may be subject to over-fitting effects, because all reference samples are taken into account explicitly, cf. Fig. 1(a).

These particular difficulties of kNN schemes motivated the idea to replace the complete set of exemplars $I\!D$ by a few representatives already in [30]. Learning Vector Quantization (LVQ) as a principled approach to the identification of suitable prototypes $\mathbf{w}^k \in I\!R^N$ $(k = 1, 2, \ldots K)$ was suggested by Kohonen [35,37]. The prototypes carry fixed labels $y^k = y(\mathbf{w}^k)$ indicating which class they represent. Obviously, the LVQ system should comprise at least one prototype per class.

Originally, LVQ was motivated as an approximate realization of a Bayes classifier with the prototypes serving as a robust, simplified representation of class-conditional densities [35,37,67]. Ideally, prototypes constitute typical representatives of the classes, see [26] for a detailed discussion of this property. Recent reviews of prototype based systems in general and LVQ in particular can be found in [11,41,53,67].

A Nearest Prototype Classifier (NPC) assigns any feature vector $\mathbf{x}$ to the class $y^* = y(\mathbf{w}^*)$ of the closest prototype $\mathbf{w}^*(\mathbf{x})$, or $\mathbf{w}^*$ for short, which satisfies

$$d(\mathbf{w}^*, \mathbf{x}) \leq d(\mathbf{w}^j, \mathbf{x}) \text{ for } j = 1, 2, \ldots K. \tag{2}$$

Assuming that meaningful prototype positions have been determined from a given data set $I\!D$, an NPC scheme based on Euclidean distance also implements piece-wise linear class boundaries. However, since usually $K \ll P$, these are much smoother than in an NN or kNN scheme and the resulting classifier is less specific to the training data. Moreover, the NPC requires only the computation and ranking of K distances $d(\mathbf{w}^j, \mathbf{x})$. Figure 1(b) illustrates the NPC scheme as parameterized by a few prototypes and employing Euclidean distance for the same data set as shown in panel (a).

In binary problems with classes A and B, a bias can be introduced by modifying the NPC scheme: A data point $\mathbf{x}$ is assigned to class A if

$$d(\mathbf{w}^A, \mathbf{x}) \le d(\mathbf{w}^B, \mathbf{x}) + \Theta \tag{3}$$

and to class B, else. Here, $\mathbf{w}^A$ and $\mathbf{w}^B$ denote the closest prototypes carrying label A or B, respectively. The threshold Θ can be varied from large negative to large positive values, yielding true positive rate (sensitivity) and false positive rate (1-specificity) as functions of Θ. Hence, the full Receiver Operator Characteristics (ROC) can be determined [22].

2.2   Learning Vector Quantizaton

A variety of schemes have been suggested for the iterative identification of LVQ prototypes from a given dataset. Kohonen's basic LVQ1 algorithm [35] already comprises the essential ingredients of most modifications which were suggested later. It is conceptually very similar to unsupervised competitive learning [14] but takes class membership information into account, explicitly.

Upon presentation of a single feature vector $\mathbf{x}^\mu$ with class label $y^\mu = y(\mathbf{x}^\mu)$, the currently closest prototype, i.e. the so-called *winner* $\mathbf{w}^* = \mathbf{w}^*(\mathbf{x}^\mu)$ is identified according to condition (2). The Winner-Takes-All (WTA) update of LVQ1 leaves all other prototypes unchanged:

$$\mathbf{w}^* \leftarrow \mathbf{w}^* + \eta_w\, \Psi(y^*, y^\mu)\, (\mathbf{x}^\mu - \mathbf{w}^*) \quad \text{with} \quad \Psi(y, \tilde{y}) = \begin{cases} +1 \text{ if } y = \tilde{y} \\ -1 \text{ else.} \end{cases} \tag{4}$$

Hence, the winning prototype is moved even closer to $\mathbf{x}^\mu$ if both carry the same class label: $y^* = y^\mu \Rightarrow \Psi = +1$. If the prototype is meant to represent a different class, it is moved further away ($\Psi = -1$) from the feature vector. The learning rate η_w controls the step size of the prototype updates.

All examples in $\mathbb{D}$ are presented repeatedly, for instance in random sequential order. A possible initialization is to set prototypes identical to randomly selected feature vectors from their class or close to the class-conditional means.

Several modifications of the basic scheme have been considered in the literature, aiming at better generalization ability or convergence properties, see [7,36,53] for examples and further references.

LVQ1 and many other modifications cannot be formulated as the optimization of a suitable objective function in a straightforward way [59]. However, several cost function based LVQ schemes have been proposed in the literature [58,59,67]. A popular example is the so-called Generalized Learning Vector Quantization (GLVQ) as introduced by Sato and Yamada [59]. The suggested cost function is given as a sum over all examples in $\mathbb{D}$:

$$E = \sum_{\mu=1}^{P} \Phi(e^\mu) \quad \text{with} \quad e^\mu = \frac{d(\mathbf{w}^J, \mathbf{x}^\mu) - d(\mathbf{w}^K, \mathbf{x}^\mu)}{d(\mathbf{w}^J, \mathbf{x}^\mu) + d(\mathbf{w}^K, \mathbf{x}^\mu)}. \tag{5}$$

For a given $\mathbf{x}^\mu$, $\mathbf{w}^J$ represents the *closest correct* prototype carrying the correct label $y(\mathbf{w}^J) = y^\mu$ and $\mathbf{w}^K$ is the *closest incorrect* prototype with $y(\mathbf{w}^K) \neq y^\mu$, respectively. A monotonically increasing function $\Phi(e^\mu)$ specifies the contribution of a given example in dependence of the respective distances $d(\mathbf{w}^J, \mathbf{x}^\mu)$ and $d(\mathbf{w}^K, \mathbf{x}^\mu)$. Frequent choices are the identity $\Phi(e^\mu) = e^\mu$ and the sigmoidal $\Phi(e^\mu) = 1/[1 + exp(-\gamma\, e^\mu)]$ where $\gamma > 0$ controls the *steepness* [59]. Note that e^μ in Eq. (5) satisfies $-1 \leq e^\mu \leq 1$. The misclassification of a particular sample is indicated by $e^\mu > 0$, while negative e^μ correspond to correctly classified training data. As a consequence, the cost function can be interpreted as to approximate the number of misclassified samples for large γ, i.e. for *steep* Φ.

Since E is differentiable with respect to the prototype components, gradient based methods can be used to minimize the objective function for a given data set in the training phase. The popular *stochastic gradient descent* (SGD) is based on the repeated, random sequential presentation of single examples [14,17,31,56].

The SGD updates of the correct and incorrect winner for a given example $\{\mathbf{x}, y(\mathbf{x})\}$ read

$$
\mathbf{w}^J \leftarrow \mathbf{w}^J - \eta_w \frac{\partial}{\partial \mathbf{w}^J} \Phi(e) = \mathbf{w}^J - \eta_w\, \Phi'(e) \frac{2d_K}{(d_J + d_K)^2} \frac{\partial d_J}{\partial \mathbf{w}^J},
$$
$$
\mathbf{w}^K \leftarrow \mathbf{w}^K - \eta_w \frac{\partial}{\partial \mathbf{w}^K} \Phi(e) = \mathbf{w}^K + \eta_w\, \Phi'(e) \frac{2d_J}{(d_J + d_K)^2} \frac{\partial d_K}{\partial \mathbf{w}^K},
$$
$$(6)$$

where the abbreviation $d_L = d(\mathbf{w}^L, \mathbf{x})$ is used. For the squared Euclidean distance we have $\partial d_L / \partial \mathbf{w}^L = -2(\mathbf{x} - \mathbf{w}^L)$. Hence, the displacement of the correct winner is along $+(\mathbf{x} - \mathbf{w}^J)$ and the update of the incorrect winner is along $-(\mathbf{x} - \mathbf{w}^K)$, very similar to the attraction and repulsion in LVQ1. However, in GLVQ, both winners are updated simultaneously.

Theoretical studies of stochastic gradient descent suggest the use of time-dependent learning rates η_w following suitable schedules in order to achieve convergent behavior of the training process, see [17,56]. for mathematical conditions and example schedules. Alternatively, automated procedures can be employed which adapt the learning rate in the course of training, see for instance [34,65]. Methods for adaptive step size control have also been devised for batch gradient versions of GLVQ, employing the full gradient in each step, see e.g. [40,54].

Alternative cost functions have been considered for the training of LVQ systems, see, for instance, [57,58] for a likelihood based approach. Other objective functions focus on the generative aspect of LVQ [26], or aim at the optimization of the classifier's ROC [68].

2.3 Alternative Distances

Although very popular, the use of the standard Euclidean distance is frequently not further justified. It can even lead to inferior performance compared with problem specific dissimilarity measures which might, for instance, take domain knowledge into account.

A large variety of meaningful measures can be considered to quantify the dissimilarity of N-dim. vectors. Here, we mention only briefly a few important alternatives to Euclidean metrics. A more detailed discussion and further examples can be found in [9,11,29], see also references therein.

The family of Minkowski distances of the form

$$d_p(\mathbf{x}, \mathbf{y}) = \left(\sum_{j=1}^{N} |x_j - y_j|^p \right)^{1/p} \quad \text{for} \quad \mathbf{x}, \mathbf{y} \in \mathbb{R}^N \tag{7}$$

provides an important set of alternatives [39]. They fulfill metric properties (for $p \geq 1$) and Euclidean distance is recovered with $p = 2$. Employing Minkowski distances with $p \neq 2$ has proven advantageous in several practical applications, see for instance [4,25,69].

A different class of more general measures is based on the observation that the Euclidean distance can be written as

$$d_2(\mathbf{x}, \mathbf{y}) = [(\mathbf{x} \cdot \mathbf{x}) - 2\mathbf{x} \cdot \mathbf{y} + (\mathbf{y} \cdot \mathbf{y})]^{1/2}. \tag{8}$$

Replacing inner products of the form $\mathbf{a} \cdot \mathbf{b} = \sum_j a_j b_j$ by a suitable kernel function $\kappa(\mathbf{a}, \mathbf{b})$, one obtains so-called kernelized distances [63,64]. In analogy to the *kernel-trick* used in the Support Vector Machine [64], kernelized distances can be used to implicitly transform non-separable complex data to simpler problems in a higher-dimensional space, see [60] for a discussion in the context of GLVQ.

A very popular dissimilarity measure that takes statistical properties of the data into account explicitly, was suggested very early by Mahalanobis [42]. The *point-wise* version

$$d_M(\mathbf{x}, \mathbf{y}) = \left[(\mathbf{x} - \mathbf{y})^{\top} C^{-1} (\mathbf{x} - \mathbf{y}) \right]^{1/2} \tag{9}$$

employs the (empirical) covariance matrix C of the data set for the comparison of two particular feature vectors. The Mahalonobis distance is widely used in the context of the unsupervised and supervised analysis of given data sets, see [55] for a more detailed discussion.

As a last example we mention statistical divergences which can be used when observations are described in terms of densities or histograms. For instance, text can be characterized by *word counts* while color histograms are often used to summarize properties of images. In such cases, the comparison of sample data amounts to evaluating the dissimilarity of histograms. A variety of statistical divergences is suitable for this task [20]. The non-symmetric Kullback-Leibler divergence [55] constitutes a well-known measure for the comparison of densities. An example of a symmetric dissimilarity is the so-called Cauchy-Schwarz divergence [20]:

$$d_{CS}(\mathbf{x}, \mathbf{y}) = 1/2 \log \left[(\mathbf{x} \cdot \mathbf{x})(\mathbf{y} \cdot \mathbf{y}) \right] - \log \left[\mathbf{x} \cdot \mathbf{y} \right]. \tag{10}$$

It can be interpreted as a special case of more general γ-divergences, see [20,50].

In LVQ, meaningful dissimilarities do not have to satisfy metric properties, necessarily. Unlike the kNN approach, LVQ classification does not rely on the pair-wise comparison of data points. A non-symmetric measure $d(\mathbf{w}, \mathbf{x}) \neq d(\mathbf{x}, \mathbf{w})$ can be employed for the comparison of prototypes and data points as long as one version is used consistently in the winner identification, update steps, and the actual classification after training [50].

In cost function based GLVQ, cf. Eq. (5), it is straightforward to replace the squared Euclidean by more general, suitable differentiable measures $d(\mathbf{w}, \mathbf{x})$. Similarly, LVQ1-like updates can be devised by replacing the term $(\mathbf{w} - \mathbf{x})$ in Eq. (4) by $1/2 \, \partial d(\mathbf{w}, \mathbf{x})/\partial \mathbf{w}$. Obviously, the winner identification has to make use of the same distance measure in order to be consistent with the update.

It is also possible to extend gradient-based LVQ to non-differentiable distance measures like the *Manhattan distance* with $p = 1$ in Eq. (7), if differentiable approximations are available [39]. Furthermore, the concepts of LVQ can be transferred to more general settings, where data sets do not comprise real-valued feature vectors in an N-dimensional Euclidean space [41]. Methods for classification problems where only pair-wise dissimilarity information is available, can be found in [27,52], for instance.

2.4 Adaptive Distances and Relevance Learning

The choice of a suitable distance measures constitutes a key step in the design of a prototype-based classifier. It usually requires domain knowledge and insight into the problem at hand. In this context, Relevance Learning constitutes a very elegant and powerful conceptual extension of distance based classification. The idea is to fix only the basic form of the dissimilarity a priori and optimize its parameters in the training phase.

2.5 Generalized Matrix Relevance Learning

As an important example of this strategy we consider here the replacement of standard Euclidean distance by the more general quadratic form

$$d_\Lambda(\mathbf{x}, \mathbf{w}) = (\mathbf{x} - \mathbf{w})^\top \Lambda \, (\mathbf{x} - \mathbf{w}) = \sum_{i,j=1}^{N} (x_i - w_i) \, \Lambda_{ij} \, (x_j - w_j). \qquad (11)$$

While the measure is formally reminiscent of the Mahalonobis distance defined in Eq. (9), it is important to note that Λ cannot be directly computed from the data. On the contrary, its elements are considered adaptive parameters in the training process as outlined below.

Note that Euclidean distance is recovered by setting Λ proportional to the N-dim. identity matrix. A restriction to diagonal matrices Λ corresponds to the original formulation of relevance LVQ, which was introduced as RLVQ or GRLVQ in [16] and [28] respectively. There, each feature is weighted by a single adaptive factor in the distance measure.

Measures of the form (11) have been employed in various classification schemes [15,32,70,71]. Here we focus on the so-called Generalized Matrix Relevance LVQ (GMLVQ), which was introduced and extended in [18,61,62]. Applications from the biomedical and other domains are discussed in Sect. 3.

As a minimal requirement, $d_\Lambda(\mathbf{x}, \mathbf{w}) \geq 0$ should hold true for all $\mathbf{x}, \mathbf{w} \in I\!\!R^N$. This can be guaranteed by assuming a re-parameterization of the form

$$\Lambda = \Omega^\top \Omega, \quad \text{i.e.} \quad d_\Lambda(\mathbf{x}, \mathbf{w}) = [\Omega\,(\mathbf{x} - \mathbf{w})]^2 \tag{12}$$

with the auxiliary matrix $\Omega \in I\!\!R^{M \times N}$. It also implies the symmetries $\Lambda_{ij} = \Lambda_{ji}$ and $d_\Lambda(\mathbf{x}, \mathbf{w}) = d_\Lambda(\mathbf{w}, \mathbf{x})$. Frequently, a normalization $\sum_{ii} \Lambda_{ii} = \sum_{ij} \Omega_{ij}^2 = 1$ is imposed in order to avoid numerical problems.

According to Eq. (12), d_Λ corresponds to conventional Euclidean distance after a linear transformation of all data and prototypes. The transformation matrix can be $(M \times N)$-dimensional, in general, where $M < N$ corresponds to a low-dimensional intrinsic representation of the original feature vectors. Note that, even for $M = N$, the matrix Λ can become singular and d_Λ is only a *pseudo-metric* in $I\!\!R^N$: for instance, $d_\Lambda(\mathbf{x}, \mathbf{y}) = 0$ is possible for $\mathbf{x} \neq \mathbf{y}$.

In the training process, all elements of the matrix Ω are considered adaptive quantities. From Eq. (12) we obtain the derivatives

$$\frac{\partial d_\Lambda(\mathbf{w}, \mathbf{x})}{\partial \mathbf{w}} = \Omega^\top \Omega\,(\mathbf{w} - \mathbf{x}), \qquad \frac{\partial d_\Lambda(\mathbf{w}, \mathbf{x})}{\partial \Omega} = \Omega\,(\mathbf{w} - \mathbf{x})(\mathbf{w} - \mathbf{x})^\top \tag{13}$$

which can be used to construct heuristic updates along the lines of LVQ1 [8,11,41]. From the GLVQ cost function, cf. Eq. (5), one obtains the matrix update

$$\Omega \leftarrow \Omega - \eta_\Omega\,\Phi'(e)\left(\frac{2 d_\Lambda^K}{(d_\Lambda^J + d_\Lambda^K)^2}\, \frac{\partial\,d_\Lambda(\mathbf{w}^J, \mathbf{x})}{\partial \Omega} - \frac{2 d_\Lambda^J}{(d_J + d_\Lambda^K)^2}\, \frac{\partial\,d_\Lambda(\mathbf{w}^K, \mathbf{x})}{\partial \Omega} \right)$$
$$\tag{14}$$

which can be followed by a normalization step achieving $\sum_{ij} \Omega^2 = 1$. Prototypes are updated as given in Eq. (6) with the gradient terms replaced according to Eq. (13). The matrix learning rate is frequently chosen smaller than that of the prototype updates: $\eta_\Omega < \eta_w$, details can be found in [18,61]. The matrix $\Omega \in I\!\!R^{M \times N}$ can be initialized by, for instance, drawing independent random elements or by setting it proportional to the N-dim. identity matrix for $M = N$.

In the measure (11), the diagonal elements of Λ quantify the weight of single features in the distance. The inspection of the relevance matrix can provide valuable insights into the structure of the data set after training, examples are discussed in Sect. 3. Off-diagonal elements correspond to the contribution of pairs of features to d_Λ and their adaptation enables the system to cope with correlations and dependencies between the features. Note that this heuristic interpretation of Λ is only justified if all features are of the same order of magnitude, strictly speaking. In any given data set, this can be achieved by applying a z-score transformation, yielding zero mean and unit variance features. Alternatively, potentially different magnitudes of the features could be taken into account after training by rescaling the elements of Λ accordingly.

2.6   Related Schemes and Variants of GMLVQ

Adaptive distance measures of the form (11) have been considered in several realizations of distance based classifiers. For example, Weinberger et al. optimize a quadratic form in the context of nearest neighbor classification [70,71]. An explicit construction of a relevance matrix from a given data set is suggested and discussed in [15], while the gradient based optimization of an alternative cost function is presented in [32].

Localized versions of the distance (11) have been considered in [18,61,71]. In GMLVQ, it is possible to assign an individual relevance matrix Λ^j to each $\mathbf{w}^j$ or to devise class-wise matrices. Details and the corresponding modified update rules can be found in [18,61]. While this can enhance the classification performance significantly in complex problems, we restrict the discussion to the simplest case of one global measure of the form (11).

The GMLVQ algorithm displays an intrinsic tendency to yield singular relevance matrices which are dominated by a few eigenvectors corresponding to the leading eigenvalues. This effect has been observed empirically in real world applications and benchmark data sets, see [41,61] for examples. Moreover, a mathematical investigation of stationarity conditions explains this typical property of GMLVQ systems [8]. Very often, the effect allows for an interpretable visualization of the labeled data set in terms of projections onto two or three leading eigenvectors [5,41,61].

An explicit *rank control* can be achieved by using a rectangular ($M \times N$) matrix Ω in the re-parameterization (12), together with the incorporation of a penalty term for rank(Λ) $< M$ in the cost function [18,62]. For $M = 2$ or 3, the approach can also be used for the discriminative visualization of labelled data sets [5].

An important alternative to the intrinsic dimension reduction provided by GMLVQ is the identification of a suitable linear projection in a pre-processing step. This can be advantageous, in particular for nominally very high-dimensional data as encountered in e.g. bioinformatics, or in situations where the number of training samples P is smaller than the dimension of the feature space. Assuming that a given projection of the form

$$\mathbf{y} = \Psi \mathbf{x}, \quad \mathbf{v} = \Psi \mathbf{w} \quad \text{with} \quad \Psi \in \mathbb{R}^{M \times N} \tag{15}$$

maps N-dim. feature vectors and prototypes to their M-dim. representations we can re-write the distance measure of the form (11) as

$$(\mathbf{x} - \mathbf{w})^\top \Lambda (\mathbf{x} - \mathbf{w}) = (\mathbf{x} - \mathbf{w})^\top \Psi^\top \widetilde{\Lambda} \Psi (\mathbf{x} - \mathbf{w}) = (\mathbf{y} - \mathbf{v})^\top \widetilde{\Lambda} (\mathbf{y} - \mathbf{v}). \tag{16}$$

Hence, training and classification can be formulated in the M-dimensional space, employing prototypes $\mathbf{v}^j \in \mathbb{R}^M$ and an $M \times M$ relevance matrix $\widetilde{\Lambda}$. Moreover, the relation $\Lambda = \Psi^\top \widetilde{\Lambda} \Psi$ facilitates its interpretation in the original feature space.

This versatile framework allows to combine GMLVQ with, for instance, Principal Component Analysis (PCA) [55] or other linear projection techniques. Furthermore, it can be applied to the classification of functional data, where the components of the feature vectors represent an ordered sequence of values rather than a collection of more or less independent quantities. This is the case in, for instance, time series data or spectra obtained from organic samples, see [43] for examples and further references. The coefficients of a, for instance, polynomial approximation of observed data are typically obtained by a linear transformation of the form (15), where the rows of Ψ represent the basis functions. Hence, training can be performed in the lower-dimensional coefficient space, while the resulting classifier is still interpretable in terms of the original features [43].

3   Biomedical Applications of GMLVQ

In the following, selected bio-medical applications of the GMLVQ approach are highlighted. The example problems illustrate the flexibility of the approach and range from the direct analysis of relatively low-dim. data in steroid metabolomics (Sect. 3.1), the combination of relevance learning with dimension reduction for cytokine data (Sect. 3.2), and the application of GMLVQ to selected gene expression data in the context of tumor recurrence prediction (Sect. 3.3). A brief discussion with emphasis on the interpretability of the relevance matrix. Eventutally, further applications of GMLVQ for biomedical and life science data are briefly mentioned in Sect. 3.4.

3.1   Steroid Metabolomics in Endocrinology

A variety of disorders can affect the human endocrine system. For instance, tumors of the adrenal glands are relatively frequent and often found incidentally [3, 21]. The adrenals produce a number of steroid hormones which regulate important body functions. The differential diagnosis of malignant Adrenocortical Carcinoma (ACC) vs. benign Adenoma (ACA) based on non-invasive methods constitutes a highly relevant diagnostic challenge [21]. In [3], Arlt et al. explore the possibility to detect malignancy on the basis of the patient's steroid excretion pattern obtained from 24 h urine samples by means of gas chromatography/mass spectrometry (GC/MS).

The analysis of data comprising the excretion of 32 steroids and steroid metabolites was presented in [3, 13]: A data set representing a study population of 102 ACA and 45 ACC samples was analysed by means of training a GMLVQ system with one prototype per class and a single, global relevance matrix $\Lambda \in \mathbb{R}^{32 \times 32}$. In a pre-processing step, excretion values were log-transformed and in every individual training process a z-score transformation was applied.

In order to estimate the classification performance with respect to novel data representing patients with unknown diagnosis, random splits of the data set were considered: about 90% of the samples were used for training, while 10% served as a validation set. Results were obtained on average over 1000 randomized splits, yielding, for instance the threshold-averaged ROC [22], see Eq. (3).

A comparison of three scenarios provides evidence for the beneficial effect of relevance learning: When applying Euclidean GLVQ, the classifier achieves an ROC with an *area under the curve* of $AUC \approx 0.87$, see Fig. 2(a). The consideration of an adaptive diagonal relevance matrix, corresponding to GRLVQ [28], yields an improved performance with $AUC \approx 0.93$. The GMLVQ approach, cf. Sect. 2.5, with a fully adaptive relevance matrix achieves an AUC of about

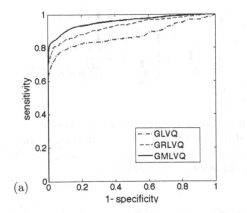
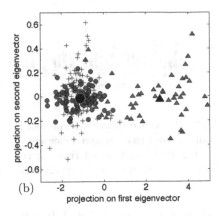

Fig. 2. Detection of malignancy in adrenocortical tumors, see Sect. 3.1. **Panel (a):** Test set ROC as obtained in the randomized validation procedure by applying GLVQ with Euclidean distance (dash-dotted line), GRLVQ with diagonal Λ (dashed) and GMLVQ with a full relevance matrix (solid). **Panel (b):** Visualization of the data set based on the GMLVQ analysis in terms of the projection of steroid profiles on the leading eigenvectors of Λ. Circles correspond to patients with benign ACA while triangles mark malignant ACC. Prototypes are marked by larger symbols. In addition, healthy controls (not used in the analysis) are displayed as crosses.

0.97. In the latter case, a working point with equal sensitivity and specificity of 0.90 can be selected by proper choice of the threshold Θ in Eq. (3). As reported in [3], the GMLVQ system outperformed alternative classifiers of comparable complexity.

The resulting relevance matrix Λ turned out to be dominated by the leading eigenvector corresponding to its largest eigenvalue; subsequent eigenvalues are found to be significantly smaller. As discussed above, this property can be exploited for the discriminative visualization of the data set and prototypes, see Fig. 2(b). The figure displays, in addition, a set of feature vectors representing healthy controls, which were not explicitly considered in the training process. Reassuringly, control samples cluster close to the ACA prototype and appear clearly separated from the malignant ACC.

By inspecting the relevance matrix of the trained system, further insight into the problem and data set can be achieved. Figure 3(a) displays the diagonal elements of Λ on average over 1000 randomized training runs. Subsets of markers can be identified, which are consistently rated as particularly important for the classification. For instance, markers 5, 6 and 19 appear significantly more relevant than all others, see [3] for a detailed discussion from the endocrinological perspective. There, the authors suggest a panel of nine leading steroids, which could serve as a reduced marker set in a practical realization of the diagnosis tool. Figure 3(b) displays the fraction of training runs in which a single marker is rated among the nine most relevant ones, providing further support for the selection of the subset [3]. Repeating the GMLVQ training for selected subsets of

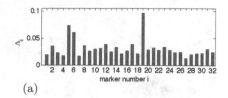

(a)   (b)

Fig. 3. Relevance of steroid markers in adrenal tumor classification, see Sect. 3.1 for details. **Panel (a):** Diagonal elements Λ_{ii} of the GMLVQ relevance matrix on average over the 1000 randomized training runs. **Panel (b):** Percentage of training runs in which a particular steroid appeared among the 9 most relevant markers.

leading markers yielded slightly inferior performance compared to the full panel of 32 markers, with $AUC \approx 0.96$ for nine steroids, and $AUC \approx 0.94$ with 3 leading markers only, see [3] for details of the analysis.

The analysis of steroid metabolomics data by means of GMLVQ and related techniques is currently explored in the context of various disorders, see [24, 38, 44] for recent examples. In the context of adrenocortical tumors, the validation of the diagnostic approach in prospective studies and the development of efficient methods for the detection of post-operative recurrence are in the center of interest [19].

3.2   Cytokine Markers in Inflammatory Diseases

Rheumatoid Arthritis (RA) constitutes an important example of chronic inflammatory disease. It is the most common form of autoimmune arthritis with symptoms ranging from stiffness and swelling of joints to, in the long term, bone erosion and joint deformity.

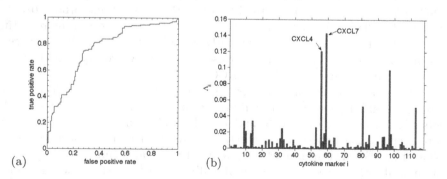

(a)   (b)

Fig. 4. GMLVQ analysis of Rheumatoid Arthritis data, see Sect. 3.2 for details. Discrimination of patients with early RA (class B) vs. resolving cases (class C). **Panel (a)** shows the ROC ($AUC \approx 0.763$) as obtained in the Leave-One-Out (from each class) validation. **Panel (b)** displays the diagonal elements of the back-transformed relevance matrix $\Lambda \in I\!R^{117 \times 117}$ on average over the validation runs.

Cytokines play an important role in the regulation of inflammatory processes. Yeo *et al.* [72] investigated the role of 117 cytokines in early stages of RA. Their mRNA expression was determined by means of PCR techniques for four different patient groups: Uninflamed healthy controls (group A, 9 samples), patients with joint inflammations that resolved within 18 months after symptom onset (group B, 9 samples), *early RA* patients developing Rheumatoid Arthritis in this period of time (group C, 17 samples), and patients with an established diagnosis of RA (group D, 12 samples).

Note that the total number of samples is small compared to the dimension $N = 117$ of the feature vectors $\mathbf{x}$ comprising log-transformed RNA expression values. Hence, standard PCA was applied to identify a suitable low-dimensional representation of the data. The analysis revealed that 95% of the variation in the data set was explained by the 21 leading principal components already. Attributing the remaining 5% mainly to *noise*, all cytokine expressions data were represented in terms of $M = 21$-dim. feature vectors corresponding to the $\mathbf{y} \in I\!\!R^M$ in Eq. (15).

GMLVQ was applied to two classification subproblems: The first addressed the discrimination of healthy controls (class A) and established RA patients (class D). While this problem does not constitute a diagnostic challenge at all, it served as a consistency check and revealed first insights into the role of cytokine markers. In the second setting, the much more difficult problem of discriminating early stage RA (class C) from resolving cases (class B) was considered.

The performances of the respective classifier systems were evaluated in a validation procedure by leaving out one sample from each class for testing and training on the remaining data. Results were reported on average over all possible test set configurations. Reassuringly, the validation set ROC obtained for the classification of A vs. D displayed almost error free performance with $AUC \approx 0.996$. The expected greater difficulty of discriminating patient groups C and D was reflected in a lower AUC of approximately 0.763, see Fig. 4(a).

It is important to note that it was not the main aim of the investigation to propose a practical diagnosis tool for the early detection of Rheumatoid Arthritis. As much as an early diagnosis would be desirable, the limited size of the study population would not provide enough supporting evidence for such a suggestion. However, the GMLVQ analysis revealed important and surprising insights into the role of cytokines. Computing the back-transformed relevance matrix $\Lambda \in I\!\!R^{117 \times 117}$ with respect to the original cytokine expression features along the lines of Eq. (16), makes possible an evaluation of their significance in the respective classification problem. Figure 4(b) displays the cytokine relevances as obtained in the discrimination of classes B and C. Two cytokines, CXCL4 and CXCL7, were identified as clearly dominating in terms of their discriminative power. A discussion of further relevant cytokines also with respect to the differences between the two classification problems can be found in [72].

The main result of the machine learning analysis triggered additional investigations by means of a direct inspection of synovial tissue samples. Careful studies employing staining techniques confirmed that CXCL4 and CXCL7 play

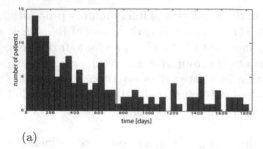
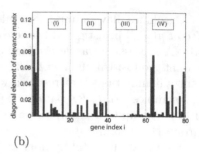

(a)   (b)

Fig. 5. Recurrence risk prediction in ccRCC, see Sect. 3.3 for details. **Panel (a):** Number of recurrences registered in the 469 patients vs. time in days. The vertical line marks a threshold of 24 months, before which 109 patients developed a recurrence. **Panel (b):** Diagonal entries of the relevance matrix with respect to the discrimination of low risk vs. high risk patients from the expression of the 80 selected genes.

an important role in the early stages of RA [72]. Significantly increased expression of CXCL4 and CXCL7 was confirmed in early RA patients compared with those with resolving arthritis or with clearly established disease. The study showed that the two cytokines co-localize, in particular, with extravascular macrophages in early stage Rheumatoid Arthritis. Implications for future research into the onset and progression of RA are also discussed in [72].

3.3 Recurrence Risk Prediction in Clear Cell Renal Cell Carcinoma

Mukherjee et al. [47] investigated the use of mRNA-Seq expression data to evaluate recurrence risk in clear cell Renal Cell Carcinoma (ccRCC). The corresponding data set is publicly available from *The Cancer Genome Atlas* (TCGA) repository [51] and is also hosted at the Broad Institute (http://gdac.broadinstitute.org). It comprises mRNA-Seq data (raw and RPKM normalized) for 20532 genes, accompanied by clinical data for survival and recurrences for 469 tumor samples. Preprocessing steps, including normalization, log-transformation, and median centering, are described in [47].

By means of an outlier analysis [1], a drastically reduced panel of 80 genes was identified for further use, see also [47] for a description of the method in this particular example. The panel consists of four different groups, each comprising 20 selected genes: In group (I), high expression can be correlated with low risk, i.e. late or no recurrence. In group (II), however, low expression is associated with low risk. Group (III) contains genes where high expression is correlated with a high risk for early recurrence, while in group (IV) low expression of the genes is an indication of high risk.

In [47], a risk index is presented, which is based on a voting scheme with respect to the 80 selected genes. Here, the focus is on the further analysis of the corresponding expression values using GMLVQ, also discussed in [47].

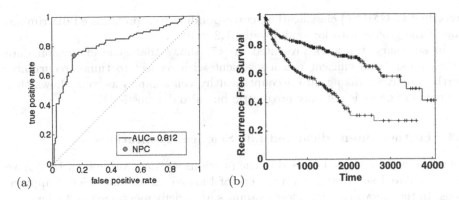

Fig. 6. Recurrence risk prediction in ccRCC, see Sect. 3.3 for details. **Panel (a):** ROC for the classification of low-risk (no or late recurrence) vs. high risk (early recurrence) as obtained in the Leave-One-Out validation of the GMLVQ classifier trained on the subset of 216 patients, cf. Sect. 3.3. The circle marks the performance of the Nearest Prototype Classifier. **Panel (b):** Kaplan-Meier plot [33] showing recurrence free survival rates in the high-risk (lower curve) and low-risk (upper curve) group as classified by the GMLVQ system applied to all 469 samples. Time is given in days.

In order to define a meaningful classification problem, two extreme groups of patients were considered: group A with poor prognosis/high risk, comprises 109 patients with recurrence within the first 24 months after the initial diagnosis. Group B corresponds to 107 patients with favorable prognosis/low risk, who did not develop tumor recurrence within 60 months after diagnosis. The frequency of recurrence times observed over five years in the complete set of 469 patients is shown in Fig. 5(a), the vertical line marks the threshold of two years after diagnosis.

A GMLVQ system with one 80-dim. prototype per class (A, B) and a global relevance matrix $\Lambda \in \mathbb{R}^{80 \times 80}$ was trained on the subset of the 216 clear-cut cases in groups A and B. Leave-One-Out validation yielded the averaged ROC shown in Fig. 6(a) with $AUC \approx 0.812$.

The diagonal elements of the averaged relevance matrix are displayed in Fig. 5(b). The results show that genes in the groups (I) and (IV) seem to be particularly discriminative and suggest that a further reduction of the gene panel should be well possible [47].

In order to further evaluate the GMLVQ classifier, it was employed to assign all 469 samples in the data set to the groups of high risk or low risk patients, respectively. In case of the 216 cases with early recurrence ($\leq$24 months) or no recurrence within 60 months, the Leave-One-Out prediction was used. For the remaining 253 patients, the GMLVQ classifier obtained from the 216 reference samples was used.

In Fig. 6 the resulting Kaplan-Meier plot [33] is shown. It displays the recurrence free survival rate of the low risk (upper) and high risk (lower) groups according to GMLVQ classification, corresponding to a pronounced discrimination of the groups with log-rank p-value 1.2×10^{-8}.

In summary, the work presented in [47] shows that gene expression data makes possible an efficient risk assessment with respect to tumor recurrence. Further analysis, taking into account healthy cell samples as well, shows that the panel of genes is not only prognostic but also diagnostic [47].

3.4   Further Bio-medical and Life Science Applications

Apart from the studies discussed in the previous sections, variants of LVQ have been employed successfully in a variety of biomedical and life science applications. In the following, a few more examples are briefly mentioned and references are provided for the interested reader.

An LVQ1-like classifier was employed for the identification of exonic vs. intronic regions in the genome of C. Elegans based on features derived from sequence data [4]. In this application, the use of the Manhattan distance in combination with heuristic relevance learning proved advantageous.

Simple LVQ1 with Euclidean distance measure was employed successfully in the inter-species prediction of protein phosphorylation in the sbv IMPROVER challenge [12]. There, the goal was to predict the effect of chemical stimuli on human lung cells, given information about the reaction of rodent cells under the same conditions.

The detection and discrimination of viral crop plant diseases, based on color and shape features derived from photographic images was studied in [50]. The authors applied divergence-based LVQ, cf. Sect. 2.3, for the comparison of feature histograms derived from Cassava plant leaf images. A comparison with alternative approaches, including GMLVQ is presented in [49].

The analysis of flow-cytometry data was considered in [6] in the context of the DREAM6/FlowCAP2 challenge [2]. For each subject, 31 markers were provided, including measures of cell size and intracellular granularity as well as 29 expression values of surface proteins for thousands of individual cells. Hand-crafted features were determined in terms of statistical moments over the entire cell population, yielding a 186-dim. representation for each patient. GMLVQ applied in this feature space yielded error-free prediction of AML in the test set [2,6].

The detection and discrimination of different Parkinsonian syndromes was addressed in [45, 46]. Three-dimensional brain images obtained by fluorodeoxyglucose positron emission tomography (FDG-PET) comprise several hundreds of thousands voxels per subject, providing information about the local glucose metabolism. An appropriate dimension reduction by *Scaled Subprofile Model with Principal Component Analysis* (SSM/PCA), yields a data set dependent, low-dimensional representation in terms of subject scores, see [45, 46] for further references. In comparison with Decision Trees and Support Vector Machines, the GMLVQ classifier displayed competitive or superior performance [46].

4   Concluding Remarks

This contribution merely serves as a starting point for studies into the application of prototype and distance based classification in the biomedical domain. It provides by no means a complete overview and focusses on the example framework of Generalized Matrix Relevance Learning Vector Quantization, which has been applied to a variety of life science datasets. The specific application examples were selected in order to demonstrate the flexibility of the approach and illustrate its interpretability.

A number of open questions and challenges deserve attention in future research – to name only a few examples: A better understanding of feature relevances should be obtained, for instance, by exploiting the approaches presented in [23]. Combined distance measures can be designed for the treatment of different sources of information in an integrative manner [48]. The analysis of functional data plays a role of increasing importance in the biomedical domain, see e.g. [43]. In general, the development of efficient methods for the analysis of biomedical data, which are at the same time powerful and transparent, constitutes a major challenge of great importance. Prototype based classifiers will continue to play a central role in this context.

Acknowledgments. The author would like to thank the collaboration partners and co-authors of the publications which are reviewed in this contribution or could be mentioned only briefly.

References

1. Aggarwal, C.: Outlier Analysis. Springer, New York (2013)
2. Aghaeepour, N., Finak, G., The FlowCAP Consortium, The DREAM Consortium*, Hoos, H., Mosmann, T., Brinkman, R., Gottardo, R., Scheuermann, R.: Critical assessment of automated flow cytometry data analysis techniques. Nat. Methods **10**(3), 228–238 (2013)
3. Arlt, W., Biehl, M., Taylor, A., Hahner, S., Libe, R., Hughes, B., Schneider, P., Smith, D., Stiekema, H., Krone, N., Porfiri, E., Opocher, G., Bertherat, J., Mantero, F., Allolio, B., Terzolo, M., Nightingale, P., Shackleton, C., Bertagna, X., Fassnacht, M., Stewart, P.: Urine steroid metabolomics as a biomarker tool for detecting malignancy in adrenal tumors. J. Clin. Endocrinol. Metab. **96**, 3775–3784 (2011)
4. Biehl, M., Breitling, R., Li, Y.: Analysis of tiling microarray data by Learning Vector Quantization and relevance learning. In: Yin, H., Tino, P., Corchado, E., Byrne, W., Yao, X. (eds.) IDEAL 2007. LNCS, vol. 4881, pp. 880–889. Springer, Heidelberg (2007). doi:10.1007/978-3-540-77226-2_88
5. Biehl, M., Bunte, K., Schleif, F.M., Schneider, P., Villmann, T.: Large margin linear discriminative visualization by matrix relevance learning. In: The 2012 International Joint Conference on Neural Networks (IJCNN), pp. 1–8, June 2012
6. Biehl, M., Bunte, K., Schneider, P.: Analysis of flow cytometry data by matrix relevance Learning Vector Quantization. PLoS ONE **8**(3), e59401 (2013). http://dx.doi.org/10.13712Fjournal.pone.0059401
7. Biehl, M., Ghosh, A., Hammer, B.: Dynamics and generalization ability of LVQ algorithms. J. Mach. Learn. Res. **8**, 323–360 (2007)

8. Biehl, M., Hammer, B., Schleif, F.M., Schneider, P., Villmann, T.: Stationarity of matrix relevance LVQ. In: 2015 International Joint Conference on Neural Networks (IJCNN), pp. 1–8, July 2015

9. Biehl, M., Hammer, B., Schneider, P., Villmann, T.: Metric learning for prototype-based classification. In: Bianchini, M., Maggini, M., Scarselli, F., Jain, L. (eds.) Advances in Neural Information Paradigms. Springer Studies in Computational Intelligence, vol. 247, pp. 183–199. Springer, Heidelberg (2010)

10. Biehl, M., Hammer, B., Verleysen, M., Villmann, T. (eds.): Similarity Based Clustering - Recent Developments and Biomedical Applications. LNAI, vol. 5400, 201 p. Springer, Heidelberg (2009)

11. Biehl, M., Hammer, B., Villmann, T.: Distance measures for prototype based classification. In: Grandinetti, L., Lippert, T., Petkov, N. (eds.) BrainComp 2013. LNCS, vol. 8603, pp. 100–116. Springer, Cham (2014). doi:10.1007/978-3-319-12084-3_9

12. Biehl, M., Sadowski, P., Bhanot, G., Bilal, E., Dayarian, A., Meyer, P., Norel, R., Rhrissorrakrai, K., Zeller, M., Hormoz, S.: Inter-species prediction of protein phosphorylation in the sbv IMPROVER species translation challenge. Bioinformatics **31**(4), 453–461 (2015)

13. Biehl, M., Schneider, P., Smith, D., Stiekema, H., Taylor, A., Hughes, B., Shackleton, C., Stewart, P., Arlt, W.: Matrix relevance LVQ in steroid metabolomics based classification of adrenal tumors. In: Verleysen, M. (ed.) 20th European Symposium on Artificial Neural Networks (ESANN 2012), pp. 423–428. d-side Publishing (2012)

14. Bishop, C.: Pattern Recognition and Machine Learning. Cambridge University Press, Cambridge (2007)

15. Boareto, M., Cesar, J., Leite, V., Caticha, N.: Supervised variational relevance learning, an analytic geometric feature selection with applications to omic data sets. IEEE/ACM Trans. Comput. Biol. Bioinform. **12**(99), 705–711 (2015)

16. Bojer, T., Hammer, B., Schunk, D., von Toschanowitz, K.T.: Relevance determination in Learning Vector Quantization. In: Verleysen, M. (ed.) European Symposium on Artificial Neural Networks, pp. 271–276 (2001)

17. Bottou, L.: Online algorithms and stochastic approximations. In: Saad, D. (ed.) Online Learning and Neural Networks, pp. 9–42. Cambridge University Press, Cambridge (1998)

18. Bunte, K., Schneider, P., Hammer, B., Schleif, F.M., Villmann, T., Biehl, M.: Limited rank matrix learning, discriminative dimension reduction, and visualization. Neural Netw. **26**, 159–173 (2012)

19. Chortis, V., Bancos, I., Sitch, A., Taylor, A., O'Neil, D., Lang, K., Quinkler, M., Terzolo, M., Manelli, M., Vassiliadi, D., Ambroziak, U., Conall Dennedy, M., Sherlock, M., Bertherat, J., Beuschlein, F., Fassnacht, M., Deeks, J., Biehl, M., Arlt, W.: Urine steroid metabolomics is a highly sensitive tool for post-operative recurrence detection in adrenocortical carcinoma. Endocrine Abstracts, vol. 41, OC1.4 (2016). doi:10.1530/endoabs.41.OC1.4

20. Cichocki, A., Zdunek, R., Phan, A., Amari, S.I.: Nonnegative Matrix and Tensor Factorizations. Wiley, Chichester (2009)

21. European Network for the Study of Adrenal Tumours: ENS@T (2002). http://www.ensat.org. Accessed 16 Mar 2017

22. Fawcett, T.: An introduction to ROC analysis. Pattern Recogn. Lett. **27**, 861–874 (2006)

23. Frenay, B., Hofmann, D., Schulz, A., Biehl, M., Hammer, B.: Valid interpretation of feature relevance for linear data mappings. In: 2014 IEEE Symposium on Computational Intelligence and Data Mining (CIDM), pp. 149–156. IEEE (2014)

24. Ghosh, S., Baranowski, E., van Veen, R., de Vries, G., Biehl, M., Arlt, W., Tino, P., Bunte, K.: Comparison of strategies to learn from imbalanced classes for computer aided diagnosis of inborn steroidogenic disorders. In: Verleysen, M. (ed.) 25th European Symposium on Artificial Neural Networks (ESANN 2017). d-side Publishing (2017, in press)

25. Golubitsky, O., Watt, S.: Distance-based classification of handwritten symbols. Int. J. Doc. Anal. Recogn. (IJDAR) **13**(2), 133–146 (2010)

26. Hammer, B., Nebel, D., Riedel, M., Villmann, T.: Generative versus discriminative prototype based classification. In: Villmann, T., Schleif, F.-M., Kaden, M., Lange, M. (eds.) Advances in Self-organizing Maps and Learning Vector Quantization. AISC, vol. 295, pp. 123–132. Springer, Cham (2014). doi:10.1007/978-3-319-07695-9_12

27. Hammer, B., Schleif, F.-M., Zhu, X.: Relational extensions of Learning Vector Quantization. In: Lu, B.-L., Zhang, L., Kwok, J. (eds.) ICONIP 2011. LNCS, vol. 7063, pp. 481–489. Springer, Heidelberg (2011). doi:10.1007/978-3-642-24958-7_56

28. Hammer, B., Villmann, T.: Generalized relevance learning vector quantization. Neural Netw. **15**(8–9), 1059–1068 (2002)

29. Hammer, B., Villmann, T.: Classification using non-standard metrics. In: Verleysen, M. (ed.) European Symposium on Artificial Neural Networks, ESANN 2005, pp. 303–316. d-side publishing (2005)

30. Hart, P.: The condensed nearest neighbor rule. IEEE Trans. Inf. Theory **14**, 515–516 (1968)

31. Hastie, T., Tibshirani, R., Friedman, J.: The Elements of Statistical Learning: Data Mining, Inference, and Prediction, 2nd edn. Springer, New York (2009)

32. Hocke, J., Martinetz, T.: Global metric learning by gradient descent. In: Wermter, S., Weber, C., Duch, W., Honkela, T., Koprinkova-Hristova, P., Magg, S., Palm, G., Villa, A.E.P. (eds.) ICANN 2014. LNCS, vol. 8681, pp. 129–135. Springer, Cham (2014). doi:10.1007/978-3-319-11179-7_17

33. Kaplan, E., Meier, P.: Nonparametric estimation from incomplete observations. J. Am. Stat. Assoc. **53**, 457–481 (1958)

34. Kingma, D.P., Ba, J.: Adam: a method for stochastic optimization. In: Proceedings of the 3rd International Conference on Learning Representations (ICLR) (2014)

35. Kohonen, T.: Learning Vector Quantization for pattern recognition. Technical report TKK-F-A601, Helsinki University of Technology, Espoo (1986)

36. Kohonen, T.: Improved versions of Learning Vector Quantization. In: International Joint Conference on Neural Networks, vol. 1, pp. 545–550 (1990)

37. Kohonen, T.: Self-organizing Maps. Springer, Heidelberg (1997)

38. Lang, K., Beuschlein, F., Biehl, M., Dietz, A., Riester, A., Hughes, B., O'Neil, D., Hahner, S., Quinkler, M., Lenders, J., Shackleton, C., Reincke, M., Arlt, W.: Urine steroid metabolomics as a diagnostic tool in primary aldosteronism. Endocrine Abstracts, vol. 38, OC1.6 (2015). doi:10.1530/endoabs.38.OC1.6

39. Lange, M., Villmann, T.: Derivatives of lp-norms and their approximations. Machine Learning Reports MLR-03-2013 (2013)

40. Biehl, M.: GMLVQ demo code (2015). http://www.cs.rug.nl/~biehl. Accessed 16 Mar 2017

41. Biehl, M., Hammer, B., Villmann, T.: Prototype-based models in machine learning. Wileys Interdisicp. Rev. (Wires) Cogn. Sci. **7**, 92–111 (2016)

42. Mahalanobis, P.: On the generalised distance in statistics. Proc. Natl. Inst. Sci. India **2**(1), 49–55 (1936)

43. Melchert, F., Seiffert, U., Biehl, M.: Functional representation of prototypes in LVQ and relevance learning. In: Merényi, E., Mendenhall, M.J., O'Driscoll, P. (eds.) Advances in Self-Organizing Maps and Learning Vector Quantization. AISC, vol. 428, pp. 317–327. Springer, Cham (2016). doi:10.1007/978-3-319-28518-4_28

44. Moolla, A., Amin, A., Hughes, B., Arlt, W., Hassan-Smith, Z., Armstrong, M., Newsome, P., Shah, T., Gaal, L.V., Verrijken, A., Francque, S., Biehl, M., Tomlinson, J.: The urinary steroid metabolome as a non-invasive tool to stage nonalcoholic fatty liver disease. Endocrine Abstracts, vol. 44, OC1.4 (2016). doi:10.1530/endoabs.44.OC1.4

45. Mudali, D., Biehl, M., Leenders, K.L., Roerdink, J.B.T.M.: LVQ and SVM classification of FDG-PET brain data. In: Merényi, E., Mendenhall, M.J., O'Driscoll, P. (eds.) Advances in Self-Organizing Maps and Learning Vector Quantization. AISC, vol. 428, pp. 205–215. Springer, Cham (2016). doi:10.1007/978-3-319-28518-4_18

46. Mudali, D., Biehl, M., Meles, S., Renken, R., Garcia-Garcia, D., Clavero, P., Arbizu, J., Obeso, J., Rodriguez-Oroz, M., Leenders, K., Roerdink, J.: Differentiating early and late stage Parkinson's disease patients from healthy controls. J. Biomed. Eng. Med. Imaging **3**, 33–43 (2016)

47. Mukherjee, G., Bhanot, G., Raines, K., Sastry, S., Doniach, S., Biehl, M.: Predicting recurrence in clear cell Renal Cell Carcinoma: analysis of TCGA data using outlier analysis and generalized matrix LVQ. In: 2016 IEEE Congress on Evolutionary Computation (CEC), pp. 656–661, July 2016

48. Mwebaze, E., Bearda, G., Biehl, M., Zühlke, D.: Combining dissimilarity measures for prototype-based classification. In: Verleysen, M. (ed.) 23rd European Symposium on Artificial Neural Networks (ESANN 2015), pp. 31–36. d-side Publishing (2015)

49. Mwebaze, E., Biehl, M.: Prototype-based classification for image analysis and its application to crop disease diagnosis. In: Merényi, E., Mendenhall, M.J., O'Driscoll, P. (eds.) Advances in Self-Organizing Maps and Learning Vector Quantization. AISC, vol. 428, pp. 329–339. Springer, Cham (2016). doi:10.1007/978-3-319-28518-4_29

50. Mwebaze, E., Schneider, P., Schleif, F.M., Aduwo, J., Quinn, J., Haase, S., Villmann, T., Biehl, M.: Divergence based classification in Learning Vector Quantization. Neural Comput. **74**(9), 1429–1435 (2011)

51. National Cancer Institute and National Human Genome Research Institute: The Cancer Genome Atlas (TCGA) Portal. http://cancergenome.nih.gov. Accessed 16 Mar 2017

52. Nebel, D., Hammer, B., Villmann, T.: A median variant of generalized Learning Vector Quantization. In: Lee, M., Hirose, A., Hou, Z.-G., Kil, R.M. (eds.) ICONIP 2013. LNCS, vol. 8227, pp. 19–26. Springer, Heidelberg (2013). doi:10.1007/978-3-642-42042-9_3

53. Nova, D., Estévez, P.: A review of Learning Vector Quantization classifiers. Neural Comput. Appl. **25**(3–4), 511–524 (2014)

54. Papari, G., Bunte, K., Biehl, M.: Waypoint averaging and step size control in learning by gradient descent (Technical report). In: Schleif, F.M., Villmann, T. (eds.) Mittweida Workshop on Computational Intelligence. MIWOCI 2011, Machine Learning Reports, volaa. MLR-2011-06, pp. 16–26. University of Bielefeld (2011)

55. Duda, R.O., Hart, P.E., Stork, D.G.: Pattern Classification. Wiley, Hoboken (2001)

56. Robbins, H., Monro, S.: A stochastic approximation method. Ann. Math. Stat. **22**, 405 (1951)

57. Seo, S., Obermayer, K.: Soft learning vector. Neural Comput. **15**, 1589–1604 (2003)

58. Seo, S., Obermayer, K.: Soft nearest prototype classification. IEEE Trans. Neural Netw. **14**, 390–398 (2003)
59. Sato, A.S., Yamada, K.: Generalized Learning Vector Quantization. In: Touretzky, D.S., Mozer, M.C., Hasselmo, M.E. (eds.) Proceedings of the Neural Information Processing Systems (NIPS), vol. 8, pp. 423–429. MIT Press, Cambridge (1996)
60. Schleif, F.-M., Villmann, T., Hammer, B., Schneider, P., Biehl, M.: Generalized derivative based kernelized Learning Vector Quantization. In: Fyfe, C., Tino, P., Charles, D., Garcia-Osorio, C., Yin, H. (eds.) IDEAL 2010. LNCS, vol. 6283, pp. 21–28. Springer, Heidelberg (2010). doi:10.1007/978-3-642-15381-5_3
61. Schneider, P., Biehl, M., Hammer, B.: Adaptive relevance matrices in Learning Vector Quantization. Neural Comput. **21**, 3532–3561 (2009)
62. Schneider, P., Bunte, K., Stiekema, H., Hammer, B., Villmann, T., Biehl, M.: Regularization in matrix relevance learning. IEEE Trans. Neural Netw. **21**, 831–840 (2010)
63. Schölkopf, B.: The kernel trick for distances. Adv. Neural Inf. Process. Syst. **13**, 301–307 (2001)
64. Shawe-Taylor, J., Cristianini, N.: Kernel Methods for Pattern Analysis, 474 p. Cambridge University Press, Cambridge (2004)
65. Schaul, T., Zhang, S., LeCun, Y.: No more pesky learning rates. JMLR: W&CP **28**, 342–351 (2013)
66. Cover, T.M., Hart, P.: Nearest neighbor pattern classification. IEEE Trans. Inf. Theory **13**, 21–27 (1967)
67. Villmann, T., Bohnsack, A., Kaden, M.: Can Learning Vector Quantization be an alternative to SVM and Deep Learning? - Recent trends and advanced variants of Learning Vector Quantization for classification learning. J. Artif. Intell. Soft Comput. Res. **7**, 65–81 (2017)
68. Villmann, T., Kaden, M., Hermann, W., Biehl, M.: Learning vector quantization classifiers for ROC-optimization. Comput. Stat., 1–22 (2016). doi:10.1007/s00180-016-0678-y
69. Villmann, T., Kästner, M., Nebel, D., Riedel, M.: ICMLA face recognition challenge - results of the team 'Computational Intelligence Mittweida'. In: Proceedings of the International Conference on Machine Learning Applications (ICMLA 2012), pp. 7–10. IEEE Computer Society Press (2012)
70. Weinberger, K., Blitzer, J., Saul, L.: Distance metric learning for large margin nearest neighbor classification. In: Weiss, Y., Schölkopf, B., Platt, J. (eds.) Advances in Neural Information Processing Systems, vol. 18, pp. 1473–1480. MIT Press, Cambridge (2006)
71. Weinberger, K., Saul, L.: Distance metric learning for large margin nearest neighbor classification. J. Mach. Learn. Res. **10**, 207–244 (2009)
72. Yeo, L., Adlard, N., Biehl, M., Juarez, M., Smallie, T., Snow, M., Buckley, C., Raza, K., Filer, A., Scheel-Toellner, D.: Expression of chemokines CXCL4 and CXCL7 by synovial macrophages defines an early stage of rheumatoid arthritis. Ann. Rheum. Dis. **75**, 763–771 (2015)

Describing the Local Structure
of Sequence Graphs

Yohei Rosen[1,2] (iD), Jordan Eizenga[2] (iD), and Benedict Paten[2(✉)]

[1] New York University School of Medicine,
550 1st Avenue, New York, NY 10016, USA
yohei.rosen@nyumc.org
[2] University of California Santa Cruz Genomics Institute,
1156 High Street, Mailstop CBSE, Santa Cruz, CA 95064, USA
benedict@soe.ucsc.edu

Abstract. Analysis of genetic variation using graph structures is an emerging paradigm of genomics. However, defining genetic sites on sequence graphs remains an open problem. Paten's invention of the *ultra-bubble* and *snarl*, special subgraphs of sequence graphs which can identified with efficient algorithms, represents important first step to segregating graphs into genetic sites. We extend the theory of ultrabubbles to a special subclass where every detail of the ultrabubble can be described in a series and parallel arrangement of genetic sites. We furthermore introduce the concept of *bundle* structures, which allows us to recognize the graph motifs created by additional combinations of variation in the graph, including but not limited to runs of abutting single nucleotide variants. We demonstrate linear-time identification of bundles in a bidirected graph. These two advances build on initial work on ultrabubbles in bidirected graphs, and define a more granular concept of genetic site.

Keywords: Sequence graphs · Genetic variants

1 Background

The concept of the genetic site underpins both classical genetics and modern genomics. From a biological perspective, a site is a position at which mutations have occurred in different samples' histories, leading to genetic variation. From an engineering perspective, a site is a subgraph with left and right endpoints where traversals by paths correspond to alleles. This is useful for indexing and querying variants in paths and for describing variants in a consistent and granular manner.

Against a linear reference, it is trivial to define sites, provided that we disallow variants spanning overlapping positions. This is clearly demonstrated by VCF structure [4]. VCF sites, consisting of any number of possible alleles, are identified by their endpoints with respect to the linear reference.

If we wish to analyze a set of variants containing structural variation, highly divergent sequences or nonlinear references structures, then a linear reference

© Springer International Publishing AG 2017
D. Figueiredo et al. (Eds.): AlCoB 2017, LNBI 10252, pp. 24–46, 2017.
DOI: 10.1007/978-3-319-58163-7_2

with only non-overlapping variants is no longer a sufficient model. Datasets with one or more of these properties are becoming more common [1,10], and sequence graphs [7] have been developed as a method of representing them. However, defining sites on graphs is considerably more difficult than on linear reference structures and the creation of methods to fully decompose sequence graphs into sites remains an unsolved problem.

2   The Challenges of Defining Sites on Graphs

On a graph-based reference, the linear reference definition of a site as a position along the reference and a set of alleles fails to work for several reasons:

1. Sequences which are at the same location in linear position may not have comparable contexts. This is a consequence of having variants which cannot be represented as edits to the linear reference but rather as edits to another variant. We illustrate this with an example from *1000 Genomes* polymorphism data, visualized using Sequence Tube Maps [2] (Fig. 1).
2. Elements of sequence may not be linearly ordered. Parallel structure of the graph (3.) is one sort of non-linearity. Graphs also allow repetitive, inverted or transposed elements of sequence. These all prevent linear ordering (Fig. 2).
3. The positions spanned by different elements of variation may partially overlap. Therefore, multiple mutually exclusive segments of sequence in a region of the graph cannot be considered to be alternates to each other at a well-defined position without having to include extraneous sequence that is shared between some but not all of the "alleles."

We can expect that the density of these graph structures will increase with increasing population sizes included in datasets (Fig. 3).

Our aim will be to recognize and fully decompose subgraphs resembling Example 1 into a notion of site, and isolate these from elements of the graph resembling Examples 2 and 3.

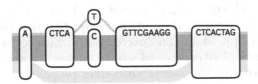

Fig. 1. The context of the single nucleotide variant shown does not exist in all variants spanning its linear position

Fig. 2. A cycle and an inversion in a graph

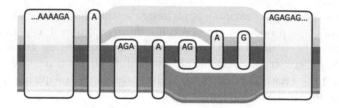

Fig. 3. Overlapping deletions, from 1000 Genomes polymorphism data

3 Mathematical Background

3.1 Directed and Bidirected Sequence Graphs

The graphs used to represent genetic information consist of labelled nodes and edges. Nodes are labelled with sequence fragments. Edges form paths whose labels spell out allowed sequences. Two types of graph are used (Fig. 4).

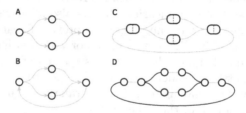

Fig. 4. (A) A directed acyclic graph (B) A cyclic directed graph (C) Graph B represented as a bidirected graph. This cycle is proper. (D) Graph C represented as a biedged graph

The more simple type is the directed graph. A directed graph (or "digraph") G consists of a set V of nodes and a set E of directed edges. A directed edge is an ordered tuple (x, y), consisting of a *head* $x \in V$ and *tail* $y \in V$. A directed path is a sequence of nodes joined by edges, followed head to tail. G is a *directed acyclic graph* (DAG) if it admits no directed path which revisits any node.

A bidirected graph G [6] consists of a set V of vertices and a set E of edges. Each vertex $v \in V$ consists of a pair of *node-sides* $\{v_{left}, v_{right}\}$ and each edge is an unordered tuple of node-sides. Bidirected graphs have the advantage of being able to represent inversion events.

We write N for the set of node-sides in the bidirected graph G. The *opposite* $\hat{n}$ of a node-side n is the other node-side at the same vertex as n.

A sequence $p = x_1, x_2, \ldots, x_k$ of node-sides is a *path* if $\forall x_i$,

1. if $x_{i-1} \neq \hat{x}_i$, then $x_{i+1} = \hat{x}_i$
2. if $x_{i-2} = \hat{x_{i-1}}$, then $\{x_{i-1}, x_i\} \in E$

3. any contiguous subsequence of p consisting of a node-side x alternating with its opposite $\hat{x}$ must either be even-numbered in length or must be a prefix or suffix of p

Informally, this means that in a path, consecutive pairs forming edges alternate with pairs of opposite node-sides or, equivalently, that paths visit both node-sides of the vertices they pass through. They can however begin or on an isolated node-side.

A bidirected graph G is *cyclic* if it admits a path visiting a node-side twice. Therefore the self-incident hairpin motif (below, right) is considered a cycle. A bidirected graph G is *properly cyclic* if it admits a path which visits a pair $\{n, \hat{n}\}$ twice in the same order (Fig. 5).

Fig. 5. (Left) A properly cyclic graph. (Right) The self-incident hairpin motif of a cyclic but not properly cyclic graph

Some publications refer to biedged graphs. These are $\{black, grey\}$-edge-colored undirected graphs, where every node is paired with precisely one other by sharing a grey edge and paths in the graph must alternate between traversing black and grey edges. Paten elaborates on this construction in [9] and shows that it is equivalent to a bidirected graph. We will restrict our language to that of bidirected graphs, recognizing that these are equivalent to biedged graphs.

Acyclic bidirected graphs are structurally equivalent to directed graphs in that

Lemma 1. *If G is a bidirected acyclic graph, there exists an isomorphic directed acyclic graph $D(G)$.*

Proof. See [9].

3.2 Bubbles, Superbubbles, Ultrabubbles and Snarls

The first use of local graph structure to identify variation was the detection of *bubbles* [13] in order to detect and remove sequencing errors from assembly graphs. Their bubble is the graph motif consisting of two paths which share a source and a sink but are disjoint between.

The general concept of bubbles was extended by Onodera et al., who defined superbubbles in directed graphs [8]. Brankovic demonstrates an $\mathcal{O}(|V| + |E|)$ algorithm to identify them [3], building off work of Sung [11].

We restate the Onodera definition, modified slightly as to be subgraph-centric rather than boundary-centric: A subgraph $S \subseteq G$ of a directed graph is a *superbubble with boundaries* (s, t) if

1. (reachability) t is reachable from s by a directed path in S
2. (matching) the set of vertices reachable from s without passing through t is equal to the set of vertices from which t is reachable without passing through s, and both are equal to S
3. (acyclicity) S is acyclic
4. (minimality) there exists no $t' \in S$ such that boundaries (s, t') fulfil 1, 2 and 3. There exists no $s' \in S$ such that (s', t) fulfil 1, 2 and 3.

To motivate our definition of a superbubble equivalent on bidirected graphs, we prove some consequences of the matching property.

Proposition 2. *Let $S \subseteq G$ be a subgraph of a directed graph. If S possesses the matching property relative to a pair (s, t), then it possesses the following three properties:*

1. *(2-node separability) Deletion of all incoming edges of s and all outgoing edges of t disconnects S from the remainder of the graph.*
2. *(tiplessness) There exist no node $n \in S\backslash\{s, t\}$ such that n has either only incoming or outgoing edges.*
3. *S is weakly connected.*

Proof. (matching $\Rightarrow$ separability) Suppose $\exists x \notin S, y \in S\backslash\{s, t\}$ such that there exists either an edge $x \to y$ or an edge $y \to x$. Suppose wlog that $\exists$ an edge $x \to y$. By matching, there exists a path $y \to \cdots \to t$ without passing through s. We can then construct the path $x \to y \to \cdots \to t$ which does not pass through s. But by matching this implies that $x \in S$, which leads to a contradiction.

The converse need not be true on directed graphs[1]. We define two structures on bidirected graphs. The first is the ultrabubble, which given Proposition 2, can be thought of as an analogue to a superbubble. The second, the snarl, is a more general object which preserves the property of 2-node separability from the larger graph without having strong guarantees on its internal structure. The following definitions are due to Paten [9]:

A connected subgraph $S \subseteq G$ of a bidirected graph G is a *snarl* (S, s, t) with boundaries (s, t), if

1. $s \neq \hat{t}$
2. (2-node separability) every path between a pair of node-sides in $x \in S, y \in G\backslash S$ contains either $s \to \hat{s}$ or $t \to \hat{t}$ as a subpath.
3. (minimality) there exists no $t' \in S$ such that boundaries (s, t') fulfil 1 and 2. There exists no $s' \in S$ such that (s', t) fulfil 1 and 2

The class of *ultrabubbles* is the subclass of snarls (S, s, t) furthermore fulfilling

4. S is acyclic
5. S contains no tips — vertices having one node-side involved in no edges

Three examples of ultrabubbles are shown below (Fig. 6).

---

[1] It is on bidirected graphs.

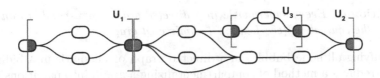

Fig. 6. Three ultrabubbles, boundaries colored blue, pink and green. These illustrate the non-overlapping property (Color figure online)

The following is important property of snarls.

Proposition 3 (Non-overlapping property). *If two distinct snarls share a vertex (node-side pair) then either they share a boundary node or one snarl is included in the other's interior.*

Proof. Let S be a snarl with boundaries s, t. Let T be another snarl, with boundaries u, v. Suppose that u ∈ S\{s, t} but v ∉ S, and s ∉ T.

Consider the set S ∩ T. It is nonempty since it contains u. Let x ∈ S ∩ T. Let y ∉ S ∩ T. Suppose that there exists a path p = x ↔ · · · ↔ y which neither passes through u nor t.

Since y ∉ S ∩ T, either y ∉ S or y ∉ T. Wlog, assume y ∉ T. Then due to the separability of T, since the path p does not pass through u, it must pass through v before leaving T to visit y. But v ∉ S so p must also pass through s before leaving S to visit v since it does not pass through t. But it must pass through v before leaving T to visit s, which leads to an impossible sequence of events. Therefore any path x ↔ · · · ↔ y for x ∈ S ∩ T, y ∉ S ∩ T must pass through either u or t. This contradicts the minimality of both S and T.

This non-overlapping property is also a nesting property. Observe that, due to Proposition 3, the relation $U \leq V$ on snarls U, V defined such that $U \leq V$ if U is entirely contained in V has the property that if $U \leq V$ and $U \leq W$, then either $V \leq W$ or $W \leq V$. Therefore the partial order on the snarls of G defined by the relation $\leq$ will always be equivalent to a tree diagram. A *bottom level* snarl is one which forms a leaf node of this tree.

Fig. 7. The nesting tree diagram of the ultrabubbles from the previous figure. U_1 and U_3 are bottom-level

The equivalent of Proposition 3 for superbubbles was stated without proof by Onodera in [8]. Our proof also constitutes a proof of the statement for superbubbles, due to the following proposition, proven by Paten in [9]:

Proposition 4. *Every superbubble in a directed graph corresponds to an ultra-bubble in the equivalent (see Lemma 1) bidirected graph.*

Identifying all superbubbles in a directed graph or all snarls in a bidirected graph introduces a method of compartmentalizing a graph into partitions whose contents are all in some sense at the same position in the graph, and for which the possible internal paths are independent of what path they continue on beyond their boundaries. We will use this concept to define sites for certain specialized classes of graphs.

4   Graphs Which Are Decomposable into Nested Simple Sites

We will extend the theory of ultrabubbles to a theory of nested sites where the structure of certain graphs can be fully described in terms of combinations of linear ordering and ultrabubble nesting relationships. This is important for

1. Identifying nested variation
2. Indexing traversals.

4.1   Traversals and Subpaths

An (s, t)-*traversal* of S is a path in S beginning with s and ending with t. An (s, s)-*traversal* and a (t, t)-*traversal* are analogously defined. Presence of an (s, s)- or (t, t)-traversal implies cyclicity. Two traversals of a snarl are *disjoint* if they are disjoint on $S \backslash \{s, t\}$.

Paten's [9] snarls and ultrabubbles are 2-node separable subgraphs whose paired boundary nodes isolate their traversals from the larger graph. We can state this with more mathematical rigor:

Claim. Consider a snarl (S, s, t) in a bidirected graph G. The set of all paths in G which contain a single (s, t)-traversal as contiguous a subpath is isomorphic to the set-theoretic product $P(s) \times Trav(s, t) \times P(t)$ consisting of the three sets

1. $P(s) := \{$paths in $G \backslash S$ terminating in $\hat{s}\}$
2. $Trav(s, t) := \{(s, t)$-traversals of $S\}$
3. $P(t) := \{$paths in $G \backslash S$ beginning with $\hat{t}\}$

The isomorphism is the function mapping $p_1 \in P(s), p_2 \in Trav(s, t), p_3 \in P(t)$ to their concatenation $p_1 p_2 p_3$.

This property is important because it allows us to express the set of all haplotypes traversing a given linear sequence of snarls in terms of combinations of alleles for which we do not need to check if certain combinations are valid.

4.2 Simple Bubbles and Nested Simple Bubbles

Definition 5. *An ultrabubble* (S, s, t) *is a simple bubble if all* (s, t)-*traversals are disjoint.*

Simple bubbles are structurally equivalent to (multiallelic) sites consisting of disjoint substitutions, insertions or deletions, with all alleles spanning the same boundaries (Fig. 8).

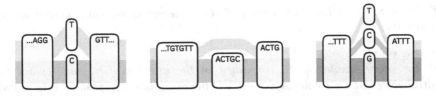

Fig. 8. Three examples of simple bubbles from the 1000 Genomes graph

Proposition 7 below demonstrates that we can identify simple bubbles in $\mathcal{O}(|V|)$ time given that we have found all snarl boundaries. Paten has shown [9] that identification of snarl boundaries is achieved in $\mathcal{O}(|E| + |V|)$ time. To find the ultrabubbles among these, note that checking for acyclicity is $\mathcal{O}(|E| + |V|)$ on account of the unbranching nature of these snarls' interiors.

Given a node-side n, write $Nb(n)$ for the set of all neighbors of n. Note that $a \in Nb(b) \Leftrightarrow b \in Nb(a)$.

Lemma 6 (Nodes in an ultrabubble are orientable with respect to the ultrabubble boundaries). *Given an ultrabubble* (S, s, t) *and given* $n \in S \backslash \{s, t\}$, *consider the set* T *of all* (s, t)-*traversals of* S *passing through* n. *Then either*

1. $\forall p \in T$, *an element of* $Nb(n)$ *precedes* n *in* p
2. $\forall p \in T$, *an element of* $Nb(n)$ *follows* n *in* p

In the former case we call n *s-sided, otherwise we call it t-sided.*

Proof. This is a corollary to Lemma 1.

Proposition 7 (Simple bubbles have unbranching interiors). *Let* (S, s, t) *be an ultrabubble. Then all traversals are disjoint iff every interior node-side has precisely one neighbor.*

Proof. ($\Rightarrow$) *Suppose that all* (s, t)-*traversals of* S *are disjoint. Suppose* $\exists$ *a node-side* $n \in S \backslash \{s, t\}$ *with multiple neighbors.*

Since n *is orientable with respect to* (s, t), *suppose, without loss of generality, that it is s-sided. Then there exist distinct paths from* s *to* n *passing through each of its neighbors. Continuing these with a path from* n^{opp} *to* t *produces two nondisjoint traversals of* S.

($\Leftarrow$) *Suppose that every interior node-side has precisely one neighbor. Suppose that there exist two distinct nondisjoint traversals of S. For no node-side to have multiple neighbors, they must coincide at every node-side, contradicting the assumption that they are not the same traversal.*

We seek to extend this simple property to more complex graph structures. We will take advantage of the nesting of nondisjoint ultrabubbles proven in Proposition 3 to define another structure in which nondisjoint traversals are easily indexed.

Definition 8. *An ultrabubble $(S, s, t) \subseteq G$ is decomposable into nested simple sites if either:*

1. *S is a simple bubble*
2. *if, for every ultrabubble S' contained in the interior of S, you replace the ultrabubble with a single edge $s - t$ whenever S' is decomposable into simple sites, then S becomes a simple bubble*

The following figure demonstrates decomposability into nested simple sites.

Proposition 9. *If an ultrabubble (U, s, t) is decomposable into nested simple sites, then the complete node sequence of any (s, t)-traversal can be determined only by specifying the path it takes inside those nested ultrabubbles within which the traversal does not visit any further nested ultrabubble.*

Proof. *Let p be a (s, t)-traversal of an ultrabubble U which is decomposable into nested simple sites. Let V be a nested ultrabubble inside U. If p traverses, V, write $p|_V$ for the traversal p restricted to V (Fig. 9).*

Suppose that $t|_V$ intersects no nested ultrabubbles within V. Then $t|_V$ is disjoint of all other traversals within V due to U begin decomposable into nested simple sites. Therefore specifying any node of $t|_V$ uniquely identifies it.

Suppose that $t|_V$ intersects some set of ultrabubbles nested within V. Since U is decomposable into nested simple sites, the nodes of $t|_V$ must be linear and disjoint of all other paths if we replace all ultrabubbles nested in V with edges joining their boundaries. Therefore specifying which ultrabubbles are crossed uniquely determines the nodes included in $t|_V$ which lie outside of the nested ultrabubbles in V.

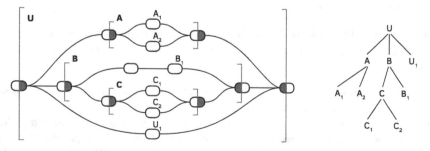

Fig. 9. Left: A nesting of four ultrabubbles. Right: The tree structure to index traversals of U implied by Proposition 9

The statement of the proposition follows from the two arguments above by induction.

Proposition 10. *An ultrabubble is decomposable into nested simple sites iff every node side is either the interior ultrabubble boundary or has precisely one neighbor.*

Proof. This can be established using Proposition 7.

This property allows $\mathcal{O}(|V| + |E|)$ evaluation of whether a graph is decomposable into nested simple sites, by arguments analogous to those for simple bubbles.

4.3   A Partial Taxonomy of Graph Notifs Which Do Not Admit Decomposition into Sites

In Sect. 4.3, we will show that we can decompose a graph into nested simple sites as defined in the previous section if it lacks a certain forbidden motif. We will begin with examples of three graph motifs, and the biological events which might produce them.

We describe some graph features which prevent decomposition into nested sites below, and the sets of mutations which might have produced them.

1. Two (or more) substitutions or deletions against a linear sequence which overlap, but not completely (Fig. 10).
2. A substitution (or deletion) which spans elements of sequence on the interior of two disjoint ultrabubbles. Addition of such an edge joining two ultrabubbles which were decomposable into nested simple sites will consolidate the two into a single ultrabubble which is not decomposable into nested simple sites (Fig. 11).
3. Two SNVs or other simple elements of variation at adjacent positions. This will be the focus of our Sect. 5.

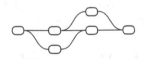

Fig. 10. Overlapping substitutions (or deletions)

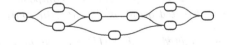

Fig. 11. An edge crossing bubble boundaries.

4.4   The Relationship Between Nested Simple Sites and Series Parallel Graphs

The structure of ultrabubbles decomposable into nested simple sites, and their tree representation (see Fig. 7) might be familiar to the graph theorist familiar with series-parallel digraphs. The fact that the digraphs equivalent to ultrabubbles form a subclass of the two-terminal series-parallel digraphs is interesting due to the computational properties of the latter class of graphs.

Definition 11. *A directed graph G is two-terminal series parallel (TTSP) with source s and sink t if either*

1. *G is the two-element graph with a single directed edge $s \to t$*
2. *There exist TTSP graphs G_1, G_2 with sources s_1, s_2 and sinks t_1, t_2 such that G is formed from G_1, G_2 by identification of s_1 with s_2 as s and identification of t_1 with t_2 as t (Parallel addition)*
3. *There exist TTSP graphs G_1, G_2 with sources s_1, s_2 and sinks t_1, t_2 such that G is formed from G_1, G_2 by identification of t_1 with s_2 (Series addition) (Fig. 12).*

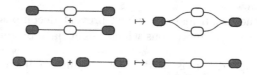

Fig. 12. Top: parallel addition. Bottom: series addition

Two terminal series parallel digraphs have a useful forbidden subgraph characterization.

Proposition 12 (From [12]). *A directed graph G is two terminal series parallel if and only if it contains no subgraph homeomorphic to the graph W shown below Proof: Refer to Valdes [12] and Duffin [5] (Fig. 13).*

Fig. 13. The W motif

Proposition 13. *If an ultrabubble (U, s, t) is decomposable into nested simple sites, then the equivalent directed graph is TTSP with source s and sink t.*

Proof. Suppose that the directed graph $D(U)$ equivalent to U (which exists by Lemma 1) contains a subgraph homeomorphic to W. Then there must be a node-side u in U with two neighbours a_1, a_2 which are the beginnings of disjoint paths p_1, p_2 ending on node-sides b_1, b_2 which are neighbours of a node-side v. By Proposition 10, u and v must be ultrabubble boundaries. Since p_1, p_2 are disjoint, *u and v must be opposing boundaries of the same ultrabubble. But the presence of a subgraph homeomorphic to W also implies that there exists a pair q_1, q_2 of disjoint paths, one from a node x to $\hat{u}$ and the other from x to v, both not passing through u or $\hat{v}$. But this is not possible since it would contradict 2-node separability of (u, v).*

We highlight the middle "Z-arm" of the W-motif in our first two examples of ultrabubbles which are not decomposable into nested simple sites.

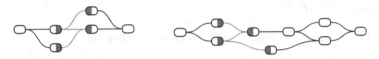

Fig. 14. Portions of the ultrabubbles 1 and 2 of Sect. 4.2, showing the nodes which project to the forbidden subgraph W

5   Abutting Variants

We wish to decompose the graph structure of sets of variants lying at adjacent positions such that there is no conserved sequence between them able to form an ultrabubble boundary. We will define a graph motif called the *balanced recombination bundle* which corresponds this graph structure, and can be rapidly detected.

We observe examples abutting single nucleotide variants (SNVs) in the 1000 Genomes polymorphism data Fig. 15. It is a reasonable hypothesis that these should become more common as the population sizes of sequencing datasets increases, since, statistically, the distribution of variation across the genome should grow less sparse as the population increases.

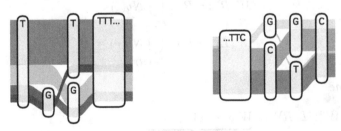

Fig. 15. Two examples of abutting SNVs in the 1000 Genomes graph

5.1   Bundles

Definition 14. *An internal chain $n_1 \to n_2 \to \cdots \to n_k$ is a sequence of node-sides such that $\forall i, 2 \le i \le k, n_i \in Nb(n_{i-1})$.*

Definition 15. *We say that a tuple (L, R) of sets of node-sides is a bundle if*

1. *(Matching)* $\forall \ell \in L, Nb(\ell) \subseteq R$ *and* $Nb(\ell) \neq \varnothing$; $\forall r \in R, Nb(r) \subseteq L$ *and* $Nb(r) \neq \varnothing$
2. *(Connectedness)* $\forall \ell \in L, r \in R$, *there exists an internal chain* $\ell \to r_1 \to \ell_1 \to \cdots \to r_k \to \ell_k \to r$ *such that* $\forall i, 1 \leq i \leq k, r_i \in R$ *and* $\ell_i \in L$

Definition 16. *We say that a tuple* (L, R) *of sets of node-sides is a balanced recombination bundle (R-bundle for short) if*

1. *(Complete matching)* $\forall \ell \in L, Nb(\ell) = R$ *and* $\forall r \in R, Nb(r) = L$
2. *(Acyclicity)* $L \cap R = \varnothing$

Lemma 17. *A balanced recombination bundle is a bundle.*

Proof. Complete matching $\Rightarrow$ matching.
   Complete matching $\Rightarrow$ connectedness by the chain $\ell \to r$ for all $\ell \in L, r \in R$.

Definition 18. *An unbalanced bundle is a bundle which is not a balanced recombination bundle. An unbalanced bundle is acyclic if* $L \cap R = \varnothing$.

Definition 19. *We say that two bundles* $(L_1, R_1), (L_2, R_2)$ *are isomorphic if either* $L_1 = L_2$ *and* $R_1 = R_2$ *or* $L_1 = R_2$ *and* $R_1 = L_2$.

We will describe a $\mathcal{O}(|V| + |E|)$ algorithm to detect and categorize bundles exhaustively for all node-sides in a bidirected graph. To establish the validity of this algorithm, we need several preliminary results:

Lemma 20. *Every* $q \in N$ *is either a tip or an element of a bundle.*

Proof. Suppose that q is not a tip. Define a function W that maps a tuple (L, R) of nonempty sets of node-sides to a tuple $W(L), W(R)$ where

$$W(R) := R \cup \bigcup_{\ell \in L} Nb(\ell)$$

$$W(L) := L \cup \bigcup_{r \in W(R)} Nb(r)$$

$\forall n \in \mathbb{N}$ *define*

$$W^n((L,R)) := \underbrace{W \circ \cdots \circ W((L,R))}_{n \ times}$$

$$W^\infty((L,R)) := W^k((L,R)) \ for \ k \ such \ that$$
$$W^{k+i}((L,R)) = W^k((L,R)) \forall i \in \mathbb{N}$$

W^∞ *exists since* W^n *is nondecreasing with respect to set inclusion and our graphs are finite. Now define* $\overline{W}(q) := W^\infty(((\{q\}, Nb(q)))$, *noting that* $Nb(q) \neq \varnothing$ *since* $\{q\}$ *is not a tip. Let us write* L_{W^∞} *and* R_{W^∞} *for the respective elements of* $\overline{W}(q)$. *We claim that* $\overline{W}(q)$ *is a bundle.*

Proof of matching: let $\ell \in L_{W\infty}, r \in R_{W\infty}$. By construction of W,

$$Nb(\ell) \subseteq W(R_{W\infty}) = R_{W\infty}$$
$$Nb(r) \subseteq W(L_{W\infty}) = L_{W\infty}$$

Proof of connectedness: let $\ell \in L_{W\infty}, r \in R_{W\infty}$. We will show that for any $r \in R_{W\infty}$, $\exists$ an internal chain $q \to r_1 \to \ell_1 \to \cdots \to r_k \to \ell_k \to r$ such that $\forall i, 1 \leq i \leq k, r_i \in R_{W\infty}$ and $\ell_i \in L_{W\infty}$.

Suppose that $r \in Nb(q)$, then we are done. Otherwise, since $r \in R_{W\infty}$, there exists some minimal $n \in \mathbb{N}$ such that $r \in$ the R-set R_{W^n} of some $W^n((\{q\}, Nb(q)))$. It is straightforward to see that we can then construct an internal chain $q \to r_0 \to \ell_1 \to r_1 \to \ldots \ell_{n-1} \to r$ such that $\forall i, 1 \leq i \leq n-1, r_i \in R_{W^i}, \ell_i \in L_{W^i}$. By an analogous argument, we can do the same for an internal chain $\ell \to \cdots \to r'$ for some $r' \in Nb(q)$. Concatenation of the first chain with the reverse of the second gives our chain $\ell \to \cdots \to r$, proving connectedness.

Proposition 21. *If $q \in L$ for a bundle (L, R), then $(L, R) = \overline{W}(q)$.*

Proof. Suppose that $\overline{W}(q) \neq (L, R)$. Then either $L \neq L_{W\infty}$ or $R \neq R_{W\infty}$. First, suppose the latter. Suppose that $\exists r \in R$ such that $r \notin R_{W\infty}$. Since (L, R) is a bundle, we know that there is an internal chain $q \to r_0 \to \ell_1 \to r_1 \to \cdots \to r_k \to \ell_k \to r$ with all $r_i \in R$, $\ell_i \in L$. But, using the same shorthand as before, it is also evident that $r_i \in R_{W^i}, \ell_i \in L_{W^i} \forall i, 1 \leq i \leq k$. But since $\ell_k \in Nb(r)$, we can deduce that $r \in R_{W^{k+1}}$, which leads to a contradiction since $r \notin R_{W\infty}$.

Suppose otherwise that $\exists r \in R_{W\infty}$ such that $r \notin R$. Consider an internal chain $c = q \to r_0 \to \ell_1 \to r_1 \to \cdots \to r_k \to \ell_k \to r$ fulfilling the conditions needed to prove connectedness of $\overline{W}(q)$. Note that $q \in L$ and by matching $r_0 \in Nb(q)$. But $r \notin R$, which leads to a contradiction since it means that there must exist two consecutive members somewhere in the chain c which cannot be neighbors.

We say that a node-side n is *involved in* a bundle (L, R) if $n \in L$ or $n \in R$.

Corollary 22 (To Proposition 21). *Every non-tip node-side is involved in precisely one bundle.*

5.2   An Algorithm for Bundle-Finding

The diagram in Fig. 16 demonstrates our algorithm for finding the balanced recombination bundle containing a query node-side q if it is contained in one, and discovering that it is not if it is not. The is written in pseudocode below, with an illustration following.

In order to prove that this is a valid algorithm for detection of balanced recombination bundles, we need the following lemma.

Lemma 23. *Let (L, R) be a tuple of sets of node-sides. If $\exists q \in L$ such that $\forall a \in Nb(q), \forall b \in Nb(a), Nb(b) \subseteq Nb(q)$ but $Nb(q) \subset R$, then (L, R) cannot be connected (in the sense of Definition 15).*

Algorithm 1. Balanced recombination bundle finding

Data: Node-side q
Result: Bundle containing q if it is in a balanced recombination bundle, $\varnothing$ if q
   is in an unbalanced bundle or is a tip
begin
│   **if** $Nb(q) = \varnothing$ **then** return $\varnothing$
│ $A \longleftarrow Nb(q)$
│ $B \longleftarrow Nb(R[0])$
│   **if** $A \cap B \neq \varnothing$ **then** return $\varnothing$
│   **else**
│   │   **for** $a \in A\backslash\{R[0]\}$ **do**
│   │   └   **if** $Nb(a) \neq B$ **then** return $\varnothing$
│   │   **for** $b \in B\backslash\{q\}$ **do**
│   │   └   **if** $Nb(b) \neq A$ **then** return $\varnothing$
│
│   return tuple (A, B)

Proof. Let $B = \bigcup_{a \in Nb(q)} Nb(a)$. We know that $\forall b \in B, Nb(b) \subseteq Nb(q)$. Suppose that (L, R) is connected. Choose $r \in R\backslash Nb(q)$. Then $\exists$ an internal chain $c = q \to r_1 \to \ell_1 \to \cdots \to r_k \to \ell_k \to r$ with $r_i \in R, \ell_i \in L \forall i$. Since $q \in B$, $Nb(b) \subseteq Nb(q) \forall b \in B$, and $Nb(a) \subseteq B \forall a \in Nb(q)$, it is impossible that the sequence of node-sides c is both a valid internal chain and ends with r. Therefore (L, R) cannot be connected.

Proposition 24 (Validity of Algorithm 1). *This algorithm detects all balanced recombination bundles, and rejects all unbalanced recombination bundles.*

Proof. Suppose q is involved in a balanced recombination bundle (L, R). W.l.o.g. suppose that $q \in L$. Due to the complete matching property, the set $Nb(q)$ in the algorithm is guaranteed to be equal to R. Due to the completeness property, the set $Nb(R[0])$ in the algorithm is guaranteed to be equal to L. It is evident that the algorithm directly verifies complete matching and acyclicity.

Suppose otherwise. Assuming we have eliminated all tips, which can be done in $\mathcal{O}(|V|)$ time, Lemma 20 proves that q is involved in an unbalanced bundle B. If B fails acyclicity but not complete matching, then checking that $A \cap B = \varnothing$ will correctly detect that $L \cap R \neq \varnothing$.

Otherwise, suppose that B fails complete matching. Suppose first that $Nb(q) \subset R$. We assert that $\exists a \in Nb(q)$ such that $\exists b \in Nb(a)$ such that $\exists c \in Nb(b)$ such that $c \notin Nb(q)$. This event will be detected by the second loop of the algorithm. This follows from the connectedness of B and Lemma 23.

Suppose otherwise that $Nb(q) = R$ but $\exists r \in R$ such that $Nb(r) \subset L$. Let $c \in L\backslash Nb(r)$. By matching, $\exists r' \in R$ such that $r' \in Nb(c)$. Therefore $Nb(r)$ and $Nb(r')$ will be found to be unequal in the first loop of the algorithm.

Suppose otherwise that $Nb(q) = R$, $Nb(r) = L \forall r \in R$, but $\exists \ell \in L$ such that $Nb(\ell) \subset R$. Then we will find in the second loop that $Nb(\ell) \neq Nb(q)$.

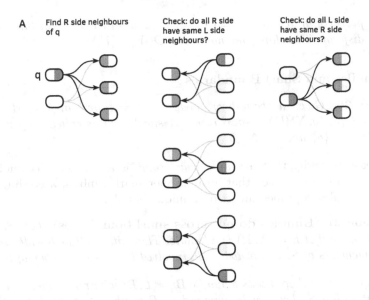

A   Find R side neighbours   Check: do all R side   Check: do all L side
         of q   have same L side   have same R side
                                    neighbours?   neighbours?

Fig. 16. Illustration of Algorithm 1 returning a positive result

Proposition 25 (Speed of Algorithm 1). *We can identify all balanced recombination bundles, all unbalanced bundles and all tips in* $\mathcal{O}(|E| + |V|)$ *time (Fig. 16).*

Proof. We depend on a neighbor index giving us $\mathcal{O}(|Nb(n)|)$ iteration across neighbors of a node-side n.

We begin by looping over all node-sides and identifying all tips, which is achieved in $\mathcal{O}(|V|)$ time. We then loop again over all remaining node-sides. At each node-side q, we run the function describe above, which, if q is involved in a balanced recombination bundle, will return the bundle $B = \overline{W}(q)$. It is evident that this function runs in $\mathcal{O}(|E_B|)$ time, seeing as it loops over each edge of B twice—once from each side—each time making an $\mathcal{O}(1)$ set inclusion query. After B is built, all nodes are marked such that they are skipped when they are encountered in the global loop. This gives overall $\mathcal{O}(|E_B|+|V_B|)$ exploration of B.

If q is involved in an unbalanced bundle $B = \overline{W}(q)$, this fact is detected by the same function in $O(|E_B|)$ time. In this case, we can find all nodes of B by performing a breadth-first search. Examination of the W-function will convince the reader that a breadth-first search will find all node-sides of B in $\mathcal{O}(|E_B| + |V_B|)$ time. We follow the same procedure of marking all these node-sides to be skipped in the global loop.

This proves that, after eliminating tips in $\mathcal{O}(|V|)$ time, we can build the set $\mathbb{B}$ of all non-isomorphic bundles B, and decide whether they are balanced recombination bundles, in time proportional to $\sum_{B \in \mathbb{B}} |E_B| + |V_B|$. But Lemma 20 and Corollary 22 tell us that $V = \{v : v \text{ is a tip}\} \cup \bigcup_{B \in \mathbb{B}} V_B$, and that all elements of this union of node-sides are disjoint. Furthermore, due to the *matching property of bundles, $E = \bigcup_{B \in \mathbb{B}} E_B$, and all elements of this union of edges are disjoint. Therefore, our method is $\mathcal{O}(|V| + |E|)$.*

5.3   Bundles and Snarl Boundaries

Definition 26. *Given a "boundary" node-side $b = s$ or t of a snarl (S, s, t), we call the tuple $(b, Nb(b))$ a snarl comb. A snarl comb is called proper if $\forall n \in Nb(b), Nb(n) = \{b\}$ and $b \notin Nb(b)$.*

It is easy to verify that a *proper snarl comb* is a balanced recombination bundle. It is also easy to see that an improper snarl comb is, according to set inclusion of tuples, a proper subset of a unique bundle.

Proposition 27 (Bundles do not cross snarl boundaries). *Let (S, s, t) be a snarl. Suppose that $B = (L, R)$ is a bundle. Then either all node-sides involved in B are members of S, or no node-side involved in B is a member of S.*

Proof. Suppose that there exists a bundle $B = (L, R)$ with node-sides both within S and not within S. Let x, y be involved in B, with $x \in S$, $y \notin S$. W.l.o.g., suppose $x \in L$, $y \in R$. This implies that there exists an internal chain $p = x \rightarrow \cdots \rightarrow y$. But then this implies that there exists $a \in S, b \notin S$ such that $a \in Nb(b)$, which would allow us to use the edge $a \rightarrow b$ to create a path violating the 2-node separability of S.

5.4   Defining Sites Using Bundles

Definition 28. *An ordered pair (B_1, B_2) of balanced recombination bundles is compatible if either*

1. *$\forall x \in R_1, \hat{x} \in L_2$, and $\forall y \in L_2, \hat{y} \in R_1$*
2. *$\exists$ a bijection $f : L_1 \longrightarrow R_2$ such that $\forall x \in R_1$, there exists a unique path $p(x)$ from $x \rightarrow \cdots \rightarrow f(x)$, and all paths $p(x)$ are disjoint.*

Definition 29. *If two recombination bundles are compatible, we define the set $p(x)$ to be a bundled simple site P.*

Claim. Consider a bundled simple site P in a graph G, lying between compatible balanced recombination bundles B_1, B_2. The set of all paths in G which contain paths $p \in P$ as contiguous subpaths is isomorphic to the set-theoretic product $P(L_1) \times P \times P(R_2)$ consisting of the three sets

1. $P(L_1) := \{$paths in $G \backslash S$ terminating in x, for some $x \in L_1\}$
2. P
3. $P(R_2) := \{$paths in $G \backslash S$ beginning with y, for some $y \in R_2\}$

under the function mapping $p_1 \in P(L_1), p \in P, p_2 \in P(R_2)$ to their concatenation.

We will call a balanced recombination bundle $B = (L, R)$ *trivial* if both L and R are singleton sets.

Definition 30. *An ultrabubble (U, s, t) is a generalized simple bubble if*

1. $(\{s\}, Nb(s))$ *and* $(Nb(t), \{t\})$ *are balanced recombination bundles*
2. *The set of all non-trivial balanced recombination bundles admits a linear ordering $X \to B_1 \to \ldots B_k \to Y$ such that X and Y are either of $(\{s\}, Nb(s))$ and $(Nb(t), \{t\})$, X is compatible with B_1, every B_i is compatible with B_{i+1}, and B_k is compatible with Y.*

Definition 31. *An ultrabubble U is decomposable into nested generalized sites if either:*

1. *It is a generalized simple bubble*
2. *When each ultrabubble (V, u, v) nested in U which is a decomposable into nested generalized sites is replaced with a single edge spanning u and v, then U is a generalized simple bubble (Fig. 17).*

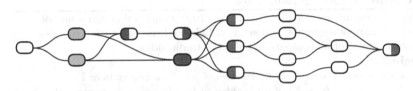

Fig. 17. An ultrabubble decomposable into nested generalized sites; some sites marked

We sketch a linear-time method of building sites from a tree diagram of nested ultrabubbles. We run Algorithms 2 and 3 starting at bottom-level nested ultrabubbles. If ultrabubble has all nontrivial balanced recombination bundles paired, then, when we evaluate the ultrabubble containing it, we represent it as a single edge from its source to sink.

In Algorithm 3, which follows below, we refer to the individual sets of node-sides forming the tuples (L, R) of a bundle as bundle-sides.

5.5 Bundles Containing Deletions

Our bundles—and therefore our sites—fail to detect the graph motifs formed by deletions spanning otherwise well-behaved variants. We define a special, well-behaved subclass of unbalanced bundle to address this (Fig. 18).

Definition 32. *A deletion bundle-pair is a tuple (L_A, R_A, L_B, R_B) such that*

1. $\forall \ell \in L_A, \forall r \in R_A, \{\ell, r\} \in E$
2. $\forall \ell \in L_A, \forall r \in R_B, \{\ell, r\} \in E$
3. $\forall \ell \in L_B, \forall r \in R_B, \{\ell, r\} \in E$
4. $\exists$ *no other edge involving any node-side $n \in L_A, L_B, R_A$ or R_B.*

Algorithm 2. Finding spans connecting bundles

Data: Ultrabubble U, and set $\mathbb{B}$ of balanced recombination bundles
Result: Set $\mathbb{P}$ of spans of unbranching sequence in U
begin

$\quad$ $\mathbb{T} \longleftarrow$ vector of all trivial bundles in $\mathbb{B}$
$\quad$ $N_{\mathbb{T}} \longleftarrow$ map $(N \rightarrow \mathbb{T})$ of node-sides to trivial bundles which contain them
$\quad$ **for** *each trivial bundle* $t = (\{t_l\}, \{t_r\}) \in \mathbb{T}$ **do**
$\quad\quad$ **if** $N_{\mathbb{T}}[\hat{t_r}]$ *is found* **then**
$\quad\quad\quad$ $(\{u_l\}, \{u_r\}) \longleftarrow N_{\mathbb{T}}[\hat{t_r}]$
$\quad\quad\quad$ replace $(\{u_l\}, \{u_r\})$ in $\mathbb{T}$ with $(\{t_l\}, \{u_r\})$
$\quad\quad\quad$ flag $(\{t_l\}, \{t_r\})$ as having been right-extended

$\quad$ $\mathbb{P} \longleftarrow \varnothing$
$\quad$ **for** $t = (\{t_l\}, \{t_r\}) \in \mathbb{T}$ **do**
$\quad\quad$ **if** t *not flagged as having been right-extended* **then** $\mathbb{P} \longleftarrow \mathbb{P} \cup \{t\}$
$\quad$ return $\mathbb{P}$

Algorithm 3. Finding compatible bundles

Data: Ultrabubble U, set $\mathbb{P}$ of node-side tuples containing endpoints of
$\quad\quad\quad$ spanning segments, and set $\mathbb{B}$ of balanced recombination bundles
Result: Set $\mathbb{C}$ of all compatible pairs of bundle-sides
begin

$\quad$ $N_{\mathbb{P}} \longleftarrow$ map $(N \rightarrow \mathbb{P})$ of node-sides of $p \in \mathbb{P}$ to elements of $\mathbb{P}$
$\quad$ $N_{\mathbb{B}} \longleftarrow$ map $(N \rightarrow \mathbb{B})$ of node-sides to bundles-sides of nontrivial R-bundles
$\quad$ $\mathbb{C} \longleftarrow \varnothing$
$\quad$ **for** *R-bundle side* $X \in \mathbb{B}$ **do**
$\quad\quad$ $x \longleftarrow X[0]$
$\quad\quad$ R-bundle side $X_{opposite} \longleftarrow \varnothing, Y \longleftarrow \varnothing$
$\quad\quad$ **if** $\hat{x}$ *found in* $N_{\mathbb{B}}$ **then**
$\quad\quad\quad$ $X_{opposite} \longleftarrow N_{\mathbb{B}}[\hat{x}]$
$\quad\quad\quad$ $Y \longleftarrow \{\hat{x}\}$

$\quad\quad$ **else if** $\hat{x}$ *found in* $N_{\mathbb{P}}$ **then**
$\quad\quad\quad$ $y \longleftarrow$ node-side of $N_{\mathbb{P}}[\hat{x}]$ which isn't $\hat{x}$
$\quad\quad\quad$ $X_{opposite} \longleftarrow N_{\mathbb{B}}[\hat{y}]$
$\quad\quad\quad$ $Y \longleftarrow \{\hat{y}\}$

$\quad\quad$ **if** $X_{opposite} \neq \varnothing$ *and* $|X_{opposite}| = |X|$ **then**
$\quad\quad\quad$ **for** $a \in X \backslash x$ **do**
$\quad\quad\quad\quad$ **if** $\hat{x}$ *found in* $N_{\mathbb{B}}$ **then** $Y \longleftarrow Y \cup \{\hat{x}\}$
$\quad\quad\quad\quad$ **else if** $\hat{x}$ *found in* $N_{\mathbb{P}}$ **then**
$\quad\quad\quad\quad\quad$ $y \longleftarrow$ node-side of $N_{\mathbb{P}}[\hat{x}]$ which isn't $\hat{x}$
$\quad\quad\quad\quad\quad$ $Y \longleftarrow Y \cup \{\hat{y}\}$

$\quad\quad\quad$ **if** $Y = X_{opposite}$ **then**
$\quad\quad\quad\quad$ $\mathbb{C} \longleftarrow \mathbb{C} \cup \{(X, X_{opposite})\}$

$\quad$ return $\mathbb{C}$

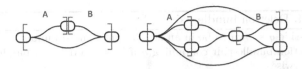

Fig. 18. Two examples of deletion bundle-pairs

These structures occur when two balanced recombination bundles on either side of some span of graph are bridged by deletions. It remains necessary to check that there is graph structure joining the nodes of R_A to L_B for this to be the case.

Algorithm 4 below will detect deletion bundle pairs from among the set of unbalanced bundles in linear time.

Proposition 33. *Given a set of acyclic unbalanced bundles, this algorithm finds those among then which are deletion bundle pairs.*

Proof. Suppose that q is involved in a deletion bundle pair (L_A, R_A, L_B, R_B). W.l.o.g, either $q \in L_A$ or $q \in L_B$.

Suppose first that $q \in L_B$: In this case, $Nb(q) = R_B$. We then know that $\forall a \in R_B, Nb(a) = L_A \cup L_B$. This will trigger the condition $L_2 = \varnothing$. The elements of $a \in L_1$ will segregate into precisely two groups: one such that $Nb(a) = R_B$— the elements $a \in L_B$, and another group such that $Nb(a) = R_A \cup R_B$—the elements $a \in L_A$. If these conditions are fulfilled, we then build R_A and R_B. It remains to verify that $\forall b \in R_A, Nb(b) = L_A$, and $\forall b \in R_B, Nb(b) = L_A \cup L_B$.

Suppose otherwise that $q \in L_A$: In this case, $Nb(q) = R_A \cup R_B$. This will trigger the condition $L_2 \neq \varnothing$ since the elements $b \in Nb(q)$ will segregate into two groups: R_A, where if $b \in R_A, Nb(b) = L_A$ and R_B, where if $b \in R_B, Nb(b) = L_A \cup L_B$. If this condition is met, then it remains to check that $\forall a \in L_A, Nb(a) = R_A \cup R_B$ and $\forall a \in L_B, Nb(a) = R_B$.

Suppose otherwise that q is not involved in a deletion bundle pair. Suppose that Algorithm 4 does not fail, returning $\varnothing$. There are two possibilities then for the nature of the unbalanced bundle (L, R) for which $q \in L$.

First, suppose the condition $L_2 = \varnothing$ was triggered. The $\exists q \in L$ such that, where $L' := \{l \in L \mid l \in Nb(a) \text{ for some } a \in Nb(q)\}$, $Nb(\ell) \subseteq Nb(q) \forall \ell \in L'$. Then by Lemma 23, $Nb(\ell) \geq Nb(q) \forall \ell \in L$. Therefore it must be that $Nb(q) = R$. Furthermore, to pass the search for R_A, there must $\exists R_A$ such that if $\ell \in L$ and $Nb(\ell) \neq R$, then $Nb(\ell) = R_A$. Furthermore, to pass the conditions of the subsequent two loops, it must be that $\forall r \in R \backslash R_A$, all $Nb(r)$ are the same, and $\forall r' \in R_A$, all $Nb(r')$ are the same. Furthermore, to pass the last condition checked, must be that $Nb(r')$ from the latter group $\subset Nb(r)$. And since $L_A := \{\ell \in L \mid Nb(\ell) = R\}$ and $L_B := \{\ell \in L \mid Nb(\ell) = R_B\}$ are such that $L_A \cap L_B = \varnothing, L_A \cup L_B = L$, these conditions all together ensure that (L, R) is a deletion bundle pair.

Otherwise, $Nb(q)$ segregates into two disjoint subsets $R_A := \{r \in Nb(q) \mid Nb(r) = L_A\}$, $R_B := \{r \in Nb(q) \mid Nb(r) = L_A \cup L_B \text{ for some}$

Algorithm 4. Deletion bundle pair finding

Data: Node-side q known to be in an acyclic unbalanced bundle

Result: Deletion bundle-pair containing q if it is in a deletion bundle-pair, $\varnothing$ otherwise

begin

 $L_A, R_A, L_B, R_B \longleftarrow \varnothing$

 $R_{temp} \longleftarrow Nb(q)$

 $L_1 \longleftarrow Nb(R_{temp}[0]), L_2 \longleftarrow \varnothing$

 for $a \in R_{temp}\backslash\{R_{temp}[0]\}$ **do**

  if $Nb(a) \neq L_1$ **then**

   if $L_2 = \varnothing$ **then** $L_2 \longleftarrow Nb(a)$

   else if $Nb(a) \neq L_2$ **then** return $\varnothing$

 if $L_2 \neq \varnothing$ **then**

  if $L_2 \subset L_1$ **then** $L_A \longleftarrow L_2$, and $L_B \longleftarrow L_1\backslash L_2$

  else if $L_1 \subset L_2$ **then** $L_A \longleftarrow L_1$, and $L_B \longleftarrow L_2\backslash L_1$

  else return $\varnothing$

  $R_{temp} \longleftarrow Nb(L_B[0])$

  for $a \in L_B\backslash\{L_B[0]\}$ **do**

   if $Nb(a) \neq R_{temp}$ **then** return $\varnothing$

  $R_{temp} \longleftarrow Nb(L_A[0])$

  for $a \in L_A\backslash\{L_A[0]\}$ **do**

   if $Nb(a) \neq R_{temp}$ **then** return $\varnothing$

  if $Nb(L_B[0]) \subset Nb(L_A[0])$ **then**

   $R_B \longleftarrow Nb(L_B[0])$

   $R_A \longleftarrow Nb(L_A[0])\backslash Nb(L_B[0])$

  else return $\varnothing$

 else

  $R_B \longleftarrow Nb(q)$

  for $a \in L_1\backslash\{q\}$ **do**

   if $Nb(a) \neq R_B$ **then**

    if $R_A = \varnothing$ **then**

     if $R_B \not\subset Nb(a)$ **then** return $\varnothing$

     else $R_A \longleftarrow Nb(a)\backslash R_B$

    else if $Nb(a) \neq R_A \cup R_B$ **then** return $\varnothing$

  if $R_A = \varnothing$ **then** return $\varnothing$

  $L_A \longleftarrow Nb(R_A[0])$

  for $a \in R_A\backslash R_A[0]$ **do**

   if $Nb(a) \neq L_A$ **then** return $\varnothing$

  $L_{temp} \longleftarrow Nb(R_B[0])$

  for $a \in R_B\backslash R_B[0]$ **do**

   if $Nb(a) \neq L_{temp}$ **then** return $\varnothing$

  if $L_A \subset L_{temp}$ **then** $L_B \longleftarrow L_{temp}\backslash L_A$

  else return $\varnothing$

 return tuple (L_A, R_A, L_B, R_B)

$L_A, L_B \subset L$ such that $L_A \cap L_B = \varnothing$. To pass further conditions, it is necessary that $\forall \ell \in L_B, Nb(\ell) = R_B$ and $\forall \ell \in L_A, Nb(\ell) = R_A \cup R_B$. It remains to show that $L_A \cup L_B = L$ and $R_A \cup R_B = R$, these can be proven by application of Lemma 23. Therefore in this case, it must also be that (L, R) is a deletion bundle pair.

Proposition 34. *This algorithm finds deletion bundles in $\mathcal{O}(|E| + |V|)$ time.*

Proof. Note that a deletion bundle-pair is a special type of unbalanced bundle. Therefore, if, given an unbalanced bundle B, we can check whether it is a deletion bundle-pair in $\mathcal{O}(|E_B| + |V_B|)$ time, by the arguments of Proposition 21, we can find all deletion bundle-pairs in $\mathcal{O}(|E| + |V|)$ time.

Inspection of the algorithm shows that, like the algorithm for identifying balanced recombination bundles, it performs two $\mathcal{O}(1)$ set-inclusion queries per edge, making it $\mathcal{O}(|E_B|)$ overall.

6 Discussion

Graph formalism has the potential to revolutionize the discourse on genetic variations by creating a model and lexicon that more fully embraces the complexity of sequence change. This is vital: the current linear genome model of a reference sequence interval and alternates is insufficient. It fails to express nested variation and can not properly describe information about the breakpoints that comprise structural variations.

The introduction, in order, of bubbles, superbubbles, ultrabubbles and snarls progressively generalizes the concept of a genetic site to accommodate more general types of variation using progressively more general graph types. In this paper we both review and build on these developments, showing how the recently introduced ultrabubbles can be furthered sub-classified using concepts from circuit theory. This expands the simple notion of proper nesting described in the original ultrabubble paper. Furthermore, we describe how we can extend the theory of ultrabubbles by generalizing ultrabubble boundaries to another sort of boundary structure—the bundle—which allows us to describe regions where variants are packed too closely to be segregated into separate ultrabubbles.

Our methods are powerful in decomposing dense collections of nested or closely packed variation into meaningful genetic sites. We anticipate that these structures will become increasingly common in the analysis of variation using graph methods, as sequencing datasets containing variation from increasing numbers of individuals become available.

Acknowledgements. Y.R. is supported by a Howard Hughes Medical Institute Medical Research Fellowship. This work was also supported by the National Human Genome Research Institute of the National Institutes of Health under Award Number 5U54HG007990 and grants from the W.M. Keck foundation and the Simons Foundation. The content is solely the responsibility of the authors and does not necessarily represent the official views of the National Institutes of Health. We thank Wolfgang Beyer for his visualizations of 1000 Genomes data in a variation graph.

References

1. 1000 Genomes Project Consortium, et al.: A global reference for human genetic variation. Nature **526**(7571), 68–74 (2015)
2. Beyer, W.: Sequence tube maps (2016). https://github.com/wolfib/sequenceTubeMap
3. Brankovic, L., Iliopoulos, C.S., Kundu, R., Mohamed, M., Pissis, S.P., Vayani, F.: Linear-time superbubble identification algorithm for genome assembly. Theor. Comput. Sci. **609**(Pt. 2), 374–383 (2016). http://www.sciencedirect.com/science/article/pii/S0304397515009147
4. Danecek, P., Auton, A., Abecasis, G., Albers, C.A., Banks, E., DePristo, M.A., Handsaker, R.E., Lunter, G., Marth, G.T., Sherry, S.T., et al.: The variant call format and vcftools. Bioinformatics **27**(15), 2156–2158 (2011)
5. Duffin, R.: Topology of series-parallel networks. J. Math. Anal. Appl. **10**(2), 303–318 (1965). http://www.sciencedirect.com/science/article/pii/0022247X65901253
6. Medvedev, P., Brudno, M.: Maximum likelihood genome assembly. J. Comput. Biol. **16**(8), 1101–1116 (2009)
7. Novak, A.M., Hickey, G., Garrison, E., Blum, S., Connelly, A., Dilthey, A., Eizenga, J., Elmohamed, M.A.S., Guthrie, S., Kahles, A., Keenan, S., Kelleher, J., Kural, D., Li, H., Lin, M.F., Miga, K., Ouyang, N., Rakocevic, G., Smuga-Otto, M., Zaranek, A.W., Durbin, R., McVean, G., Haussler, D., Paten, B.: Genome graphs. bioRxiv (2017). http://biorxiv.org/content/early/2017/01/18/101378
8. Onodera, T., Sadakane, K., Shibuya, T.: Detecting superbubbles in assembly graphs. In: Darling, A., Stoye, J. (eds.) WABI 2013. LNCS, vol. 8126, pp. 338–348. Springer, Heidelberg (2013). doi:10.1007/978-3-642-40453-5_26
9. Paten, B., Novak, A.M., Garrison, E., Hickey, G.: Superbubbles, ultrabubbles and cacti. bioRxiv (2017). http://biorxiv.org/content/early/2017/01/18/101493
10. Sudmant, P.H., Rausch, T., Gardner, E.J., Handsaker, R.E., Abyzov, A., Huddleston, J., Zhang, Y., Ye, K., Jun, G., Fritz, M.H.Y., et al.: An integrated map of structural variation in 2,504 human genomes. Nature **526**(7571), 75–81 (2015)
11. Sung, W.K., Sadakane, K., Shibuya, T., Belorkar, A., Pyrogova, I.: An o(m log m)-time algorithm for detecting superbubbles. IEEE/ACM Trans. Comput. Biol. Bioinform. **12**(4), 770–777. https://doi.org/10.1109/TCBB.2014.2385696
12. Valdes, J., Tarjan, R.E., Lawler, E.L.: The recognition of series parallel digraphs. SIAM J. Comput. **11**(2), 298–313 (1982). http://dx.doi.org/10.1137/0211023
13. Zerbino, D.R., Birney, E.: Velvet: algorithms for de novo short read assembly using de bruijn graphs. Genome Res. **18**(5), 821–829 (2008)

Graph Algorithms for Computational Biology

Mapping RNA-seq Data to a Transcript Graph via Approximate Pattern Matching to a Hypertext

Stefano Beretta, Paola Bonizzoni, Luca Denti[✉], Marco Previtali, and Raffaella Rizzi

Department of Informatics, Systems and Communication (DISCo),
University of Milan–Bicocca, Viale Sarca 336, Milan, Italy
{beretta,bonizzoni,luca.denti,marco.previtali,rizzi}@disco.unimib.it

Abstract. Graphs are the most suited data structure to summarize the transcript isoforms produced by a gene. Such graphs may be modeled by the notion of hypertext, that is a graph where nodes are texts representing the exons of the gene and edges connect consecutive exons of a transcript. Mapping reads obtained by deep transcriptome sequencing to such graphs is crucial to compare reads with an annotation of transcript isoforms and to infer novel events due to alternative splicing at the exonic level.

In this paper, we propose an algorithm based on Maximal Exact Matches that efficiently solves the approximate pattern matching of a pattern P to a hypertext H. We implement it into Splicing Graph ALigner (SGAL), a tool that performs an accurate mapping of RNA-seq reads against a graph that is a representation of annotated and potentially new transcripts of a gene. Moreover, we performed an experimental analysis to compare SGAL to a state-of-art tool for spliced alignment (STAR), and to identify novel putative alternative splicing events such as exon skipping directly from mapping reads to the graph. Such analysis shows that our tool is able to perform accurate mapping of reads to exons, with good time and space performance.

The software is freely available at https://github.com/AlgoLab/galig.

Keywords: Approximate sequence analysis · Next-generation sequencing · Alternative splicing · Graph-based alignment

1 Introduction

Typical new sequencing technologies experiments produce millions, or even billions, of reads [8]. Although the amount of transcriptomic sequencing data is smaller compared to the genomic one, the problem of aligning RNA-seq reads to a reference genome is much more complicated than that of mapping DNA-seq reads to the same reference, since RNA-seq reads reflect the biological process of alternative splicing, by which introns are removed from the DNA. Thus,

© Springer International Publishing AG 2017
D. Figueiredo et al. (Eds.): AlCoB 2017, LNBI 10252, pp. 49–61, 2017.
DOI: 10.1007/978-3-319-58163-7_3

a RNA-seq read may span two or more coding regions (also called exons) that are separated by hundreds or even thousands nucleotidic bases (introns) in the genome. The *spliced alignment* may be complicated by the repetitive structure of the genomic regions and by the short length of NGS reads. Spliced alignment is usually the first step of procedures that analyze gene expression from RNA-seq data [6,18] that try to reconcile the alignments to identify the presence of novel splicing events w.r.t. a specific gene annotation or *gene structure*, given, for example, as splicing graph [2,7].

When the annotation of the transcripts of the genes is available, the problem of transcriptome read alignment can be modeled as an approximate (since errors in the reads may occur) pattern matching to a hypertext or to a graph whose nodes are texts, that is, matching a pattern to a path of the graph allowing errors. The use of a hypertext allows to compute more accurate read alignments that, in turn, can be used to retrieve information that otherwise cannot be directly derived from the spliced alignments to a reference genome. Indeed, the hypertext models the gene structure in a concise and precise way w.r.t. a plain representation of a reference genome, and performing pattern matching to the hypertext can reveal details on how a read covers isoforms and splice junctions of a gene. The pattern matching to a hypertext problem was originally studied by Manber and Wu [12] who proposed a $\mathcal{O}(|V| + m|E| + occ \lg \lg m)$ time algorithm, where V is the set of vertices of the hypertext, E is the set of edges, m the length of the pattern, and occ the number of matches. In 2000, Amir [1] proposed a solution to this problem by giving an algorithm with time complexity $\mathcal{O}(m(n \lg m + |E|))$, where n is the total length of the hypertext. Since then, time complexity has been improved to $\mathcal{O}(m(n + |E|))$ by Navarro [13]. Thachuk [17] was the first to propose the usage of a succinct data structure to compute exact pattern matching in a hypertext, describing a very efficient algorithm with time complexity $\mathcal{O}(m \lg |\Sigma| + \gamma^2)$, where γ is number of occurrences of the node texts as substrings of the pattern and Σ is the alphabet size, but did not provide a solution to the approximated version of the problem.

In this paper, we propose an efficient algorithm for the approximate pattern matching to a hypertext problem and we implement it into a practical tool with the specific aim of mapping RNA-seq reads to a splicing graph. Our algorithm uses the same notion of Maximal Exact Matches (MEMs) of [14]. The algorithm by Ohlebusch *et al.* efficiently computes, in linear time with the length of the pattern and the number of MEMs, all the MEMs between a text and a pattern using the FM-index of the former by a backward search procedure. Our algorithm mainly consists of two steps: (i) the detection of MEMs between the pattern and the hypertext and (ii) the construction of a graph connecting MEMs that are consecutive both in the pattern and in the hypertext, allowing also errors. Finally, all the paths representing the best approximate mappings between the pattern and the hypertext are output.

There are some advantages in directly mapping reads to splicing graphs, even when the annotation is not complete. Indeed, by augmenting the hypertext with novel edges respecting the topological order of the nodes, it is possible to detect novel alternative splicing events, such as exon skipping, by directly testing the existence of approximate matchings to the added edges.

For this purpose, we implemented our approach in a tool called Splicing Graph ALigner (SGAL from now on), and we run an experimental analysis of it on RNA-seq data obtained by the sequencing of the *Toxoplasma* organism [20]. The annotation of this parasite is still incomplete and, thus, our tool has the potential to play an important role in enriching it. To evaluate the performance of SGAL, we also performed a quantitative analysis by measuring the time and space requirements of our implementation and compared it with one of the most used spliced aligner (STAR [6]). The results shows that SGAL is competitive in dealing with real RNA-seq data and offers a good scalability which is promising for future analysis involving Third-Generation Technologies data.

2   Preliminaries

Given a string S over an alphabet Σ, then $S[i]$ and $S[i,j]$ denote the ith character and the substring from the ith to the jth character of S, respectively.

A *hypertext* H is a directed graph (V, E) where each node $v \in V$ is labeled by a string T_v over the alphabet Σ. In this work, we will focus on acyclic hypertexts, that is DAGs. A path $\pi = \langle v_1, v_2, \cdots, v_n \rangle$ of H represents the string produced by the concatenation of the labels of its nodes.

We say that a string S *fully-covers* the path π if S consists of the string $T' \cdot T_{v_2} \cdot T_{v_3}, \cdots, \cdot T_{v_{n-1}} \cdot T''$ where T' is a suffix of T_{v_1} (or the entire T_{v_1}) and T'' is a prefix of T_{v_n} (or the entire T_{v_n}). We say that a string P over Σ (pattern) *matches* a path π, if there exists a string S that fully-covers π which differs from P for a limited number of errors. Let $\sigma_H = \langle v_1, v_2, \cdots, v_{|V|} \rangle$ be a topological sorting of the nodes of a hypertext $H = (V, E)$ [4, Sect. 22.4]. Then, the *Hypertext Serialization* of H w.r.t. σ_H is the string $T_H = \phi T_{v_1} \phi T_{v_2} \phi \cdots \phi T_{v_{|V|}} \phi$ obtained by concatenating the labels of the nodes taken in the same order as σ_H and interposing between them a special character ϕ, lexicographically smaller than any $c \in \Sigma$. In the following, given a position j on T_H such that $T_H[j] \neq \phi$, we denote by map(j) the node v_i of H such that $j > i + \Sigma_{k=1}^{i-1} |T_{v_k}|$ and $j < i + \Sigma_{k=1}^{i} |T_{v_k}| + 1$. Informally, map($j$) is the node v_i of H whose label T_{v_i} in T_H

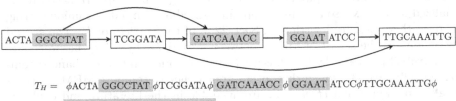

$T_H = \quad \phi\text{ACTA GGCCTAT }\phi\text{TCGGATA}\phi\text{ GATCAAACC }\phi\text{ GGAAT ATCC}\phi\text{TTGCAAATTG}\phi$

$P = \quad \text{GGCACTATGATCCAACCGGAT}$

Fig. 1. An example of a Hypertext $H = (V, E)$ with its serialization T_H is shown. Moreover, the approximate alignment of a pattern P to T_H is highlighted by shadowed boxes.

contains position j. Moreover, we denote by $\text{start}(j) = i + \Sigma_{k=1}^{i-1}|T_{v_k}| + 1$ and $\text{end}(j) = i + \Sigma_{k=1}^{i}|T_{v_k}|$ the start and the end position of the label T_{v_i} containing index j inside T_H, respectively. An example of hypertext serialization is shown in Fig. 1.

Hypertexts are elegant and straightforward representations of collections of texts. One of the most interesting problems concerning this data structure asks for finding all the occurrences of a given string in the hypertext; such problem is usually referred to as *Approximate Pattern Matching in Hypertexts* problem (APMH), and can be formalized as follows. Given a hypertext H, a pattern P, and a threshold ϵ, find all the paths of H to which a string S, having edit distance from P smaller than ϵ, matches. An example is given in Fig. 1.

Given two strings T and P of length n and m respectively, a *Maximal Exact Match (MEM)* between T and P is a triple (t, p, ℓ) such that: (i) $T[t, t + \ell - 1] = P[p, p + \ell - 1]$, (ii) $p + \ell - 1 = m$ or $t + \ell - 1 = n$ or $T[t + \ell] \neq P[p + \ell]$, and (iii) $p = 1$ or $t = 1$ or $T[t - 1] \neq P[p - 1]$. Informally, a MEM represents a common substring of length ℓ between T and P that cannot be extended in either direction without introducing a mismatch. Computing MEMs between two strings is a widely studied problem in the literature. In 2010, Ohlebusch *et al.* [14] proposed a method to efficiently compute MEMs in time linear to the length of the pattern and the number of the MEMs and in compressed space, by using enhanced compressed suffix arrays.

3   Methods

In this section we propose an algorithm to solve the APMH. Given a hypertext H, a pattern P, two integers L and K, an error threshold ϵ, and a value $C \in [0, 1]$ corresponding to the minimum coverage of P, our method can be summarized into the following steps: (1) H is serialized into the text T_H, (2) all the MEMs having a minimum length L between P and T_H are computed, (3) the MEMs are processed in order to construct the paths linking MEMs having distance up to K and covering more than $C \cdot |P|$ positions of P, and (4) all the *matching substrings* having edit distance smaller than ϵ are extracted from those paths. In the following, we consider n as the length of the serialized text T_H and m as the length of pattern P.

Since MEMs are computed between two strings, the graph H needs to be serialized into a text, so as a first step our procedure computes a unique string T_H from the topological sorting [4, Sect. 22.4] of the vertices of H. The second step computes the set M of MEMs between P and T_H by using the procedure `backwardMEM` proposed by Ohlebusch *et al.* [14], which employs enhanced compressed suffix arrays. Notice that, other approaches to compute MEMs, such as `essaMEM` by Vyverman *et al.* [19], are available in the literature and could be used in this step. The third step processes the set M of MEMs. Before describing this step we must introduce the MEM-graph $\mathcal{G} = (M, E_M)$, where the edge set E_M gives all the *K-consecutive* pairs (m_1, m_2) of MEMs. More in detail, we say that a pair of MEMs (m_1, m_2), where $m_1 = (t_1, p_1, \ell_1)$ and $m_2 = (t_2, p_2, \ell_2)$,

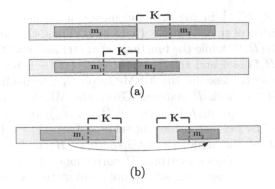

(a)

(b)

Fig. 2. K-consecutivity: in (a) an example of *Pattern-K-consecutivity* of two MEMs m_1 and m_2. In (b) an example of *Hypertext-K -consecutivity* of two MEMs m_1 and m_2.

is K-*consecutive* if it is both *Pattern-K-consecutive* and *Hypertext-K-consecutive*. Formally, (m_1, m_2) is *Pattern-K-consecutive* if: (i) $p_1 < p_2$, (ii) $p_1 + \ell_1 < p_2 + \ell_2$, and (iii) either $p_2 - p_1 - \ell_1 \leq K$ if $p_2 \geq p_1 + \ell_1$ or $p_1 + \ell_1 - p_2 \leq K$ if $p_2 < p_1 + \ell_1$. Observe that the first condition requires that the begin of m_1 is before the begin of m_2 on P, the second one requires that the end of m_1 is before the end of m_2 on P, and the third one allows a limited gap or a limited overlap between the two MEMs (see Fig. 2(a)).

Moreover, we say that the pair of MEMs (m_1, m_2) is *Hypertext-K-consecutive* in two cases based on the fact that m_1 and m_2 are in the same node of H or not. In the former situation, that is, positions t_1 and t_2 are in the same label T_{v_i} of the serialization text T_H, we say that (m_1, m_2) is *Hypertext-K-consecutive* if: (i) $t_1 < t_2$, (ii) $t_1 + \ell_1 < t_2 + \ell_2$, and (iii) either $t_2 - t_1 - \ell_1 \leq K$ if $t_2 \geq t_1 + \ell_1$ or $t_1 + \ell_1 - t_2 \leq K$ if $t_2 < t_1 + \ell_1$. Observe that these are the same conditions of the *Pattern-K-consecutivity* (see Fig. 2(a)). In the latter case, that is, positions t_1 and t_2 are in two different labels T_{v_i} and T_{v_j} of the serialization text T_H, we say that (m_1, m_2) is *Hypertext-K-consecutive* if: (i) there exists the edge (v_i, v_j) in H, (ii) $e - t_1 - \ell_1 + 1 \leq K$, and $t_2 - b \leq K$. The second condition requires that m_1 and m_2 map to two substrings sufficiently close to the right end of the label T_{v_i} and to the left end of the label T_{v_j}, respectively (see Fig. 2(b)). Notice that, the value of K should be greater than that of L in order to be able to find and possibly connect MEMs separated by more than K positions, in which no other MEM having length at least L is found.

In order to test the K-consecutivity of two MEMs m_1 and m_2 on T_H, we need to efficiently compute $v_i = \texttt{map}(t_1)$, $v_j = \texttt{map}(t_2)$, $b = \texttt{start}(t_2)$, and $e = \texttt{end}(t_1)$. Recall that, given a position t on T_H, $\texttt{map}(t)$ is the node of H whose label contains position t, while $\texttt{start}(t)$ and $\texttt{end}(t)$ are the start and the end position of the label inside T_H, respectively. To this aim, we use a bit vector B such that $B[i] = 1$ if and only if $T_H[i] = \phi$, otherwise $B[i] = 0$. In general, a bit vector B is a binary vector of length n, that is, $\forall i\, 1 \leq i \leq n, B[i] \in \{0, 1\}$, supporting in $\mathcal{O}(1)$ time the following operations: (i) $rank_b(B, i)$, returning the number of bits $b \in \{0,1\}$ in the first i elements of B, and (ii) $select_b(B,i)$, returning the position of the ith bit $b \in \{0,1\}$ in B. Thus, the function $\mathtt{map}(t)$ is computed as $rank_1(B,t)$, while the functions $\mathtt{start}(t)$ and $\mathtt{end}(t)$ are computed as $select_1(rank_1(B,t)) + 1$ and $select_1(rank_1(B,t)+1) - 1$, respectively.

Now we are able to describe the MEM-Graph $\mathcal{G}$, which allows to retrieve all the paths of H to which P matches. Given the MEM-Graph $\mathcal{G}$, let $\mathcal{P}_{\mathcal{G}} = \langle m_1 = (t_1, p_1, \ell_1), m_2 = (t_2, p_2, \ell_2), \cdots, m_n = (t_n, p_n, \ell_n) \rangle$ be a source-sink path composed of n MEMs. Let us consider the set $\mathcal{P} = \{\mathtt{map}(t) \mid t \in \{t_1, t_2, \cdots, t_n\}\}$ which is a subset of the nodes of the hypertext H, and it is possible to show that, after a topological sorting, $\mathcal{P}$ corresponds to a path of H. Let $\mathcal{P} = \{v_1, v_2, \cdots, v_q\}$ be those nodes after topological sorting. Observe that $n \geq q$ since two MEMs can refer to the same node. It is possible to prove that the MEMs of the path $\mathcal{P}_{\mathcal{G}}$ follows this order and, moreover, m_1 occurs in the label T_{v_1} of v_1 and m_n occurs in the label T_{v_q} of v_q. Let $S_{\mathcal{P}_{\mathcal{G}}} = T'T_{v_2} \cdots T_{v_{q-1}}T''$, where T' is the suffix of the label T_{v_1} starting at position t_1 and T'' is the prefix of T_{v_q} ending at position $t_n + \ell_n$, and let $S_P = P[p_1, p_n + \ell_n - 1]$ be the substring of P starting at p_1 and ending at $p_n + \ell_n - 1$. If the strings $S_{\mathcal{P}_{\mathcal{G}}}$ and S_P have $\epsilon' \leq \epsilon$ errors, then $S_{\mathcal{P}_{\mathcal{G}}}$ is a candidate *matching string*, which could correspond to the entire pattern P or a substring of it, and $\mathcal{P}$ is candidate to be a path of H to which P matches. More precisely, if $S_{\mathcal{P}_{\mathcal{G}}}$ does not match the entire pattern, then a prefix P' of length $p_1 - 1$ and a suffix P'' of length $|P| - p_n - \ell_n + 1$ of P are not part of the matching. This is due to the fact that there are no MEMs having length at least L in those two regions. In this case, to have a candidate *matching string*, the both errors between P' and the region on the hypertext before the occurrence of MEM m_1, and between P'' and the region on the hypertext after the occurrence of MEM m_n must be lower than $\epsilon - \epsilon'$, since ϵ' errors are those in the substring $S_{\mathcal{P}_{\mathcal{G}}}$ of the pattern already matching.

In any case, we are interested in matches where the discarded prefixes and suffixes are sufficiently short, that is, we want to guarantee a sufficiently high coverage of the matching region over P. For this reason, the third step of our algorithm computes from the set M of MEMs, the subgraph $\mathcal{G}' = (M', E_{M'})$ of $\mathcal{G}$ (called *reduced* MEM-Graph), containing only the source-sink paths $\langle m_1 = (t_1, p_1, \ell_1), \cdots, m_n = (t_n, p_n, \ell_n) \rangle$ having $p_1 \leq (1 - C) \cdot m + 1$ and $p_n + l_n \geq C \cdot m$ (recall that $C \in [0,1]$ is the minimum coverage of P). The rationale is that, if $p_1 > (1 - C) \cdot m + 1$, the length of the discarded prefix is likely to be too high, since we expect to have too many errors in the prefix $P[1, p_1 - 1]$. Analogously, if $p_n + l_n < C \cdot m$, the length of the discarded suffix is likely to be too high, since we expect to have too many errors in the suffix $P[p_n + l_n, m]$. For this reason, a source node $m_f = (t_f, p_f, \ell_f)$ is added to $\mathcal{G}'$ if and only if $p_f \leq (1 - C) \cdot m + 1$ while each sink node $m_e = (t_e, p_e, \ell_e)$ in $\mathcal{G}'$ such that $p_e + \ell_e \geq C \cdot m$ is labeled as *valid*. Thus, the graph $\mathcal{G}'$ is visited from *valid* sink nodes to source ones to avoid visiting the paths of $\mathcal{G}'$ which are likely to lead to a bad matching.

The reduced MEM-Graph $\mathcal{G}' = (M', E_{M'})$ is built by processing the set M of MEMs between P and T_H, by increasing values of their positions on the pattern P. To improve the efficiency, M is stored in a vector L_M of length

m such that $L_M[p]$ is the list of MEMs starting at position p of P. Initially, M' and $E_{M'}$ are set as empty. Then, the algorithm considers each position p_i on P and the list $L_M[p_i]$ of MEMs starting at p_i is retrieved. For each MEM $m_i = (t_i, p_i, \ell_i)$ in $L_M[p_i]$, if m_i is not in M' and $p_i > (1 - C) \cdot m + 1$, it is skipped. Otherwise, m_i is added to M' and the algorithm selects all the MEMs in $(L_M[p_i + \ell_i - K], \ldots, L_M[p_i + \ell_i + K])$ that are *Pattern-K-consecutive* and *Hypertext-K-consecutive* with m_i and saves them in a set M_j. After that, for each $m_j = (t_j, p_j, \ell_j) \in M_j$, the algorithm adds the node m_j to M' and the edge (m_i, m_j) to $E_{M'}$. Finally, if $p_j + l_j \geq C \cdot m$, m_j is labeled as *valid*.

In the last step of the algorithm, $\mathcal{G}'$ is visited using a Depth-First Search, starting from each sink vertex labeled as *valid*, to retrieve all the paths corresponding to candidate matchings strings and the ones that do not correspond to the entire pattern are extended as described above.

4   Experimental Analysis

We developed a tool that implements the APMH algorithm we proposed in Sect. 3 in order to perform the approximate mapping of RNA-seq data to splicing graphs (hypertexts), and we assessed its performance on a real dataset. Such tool, called SGAL, takes as input a GFF file containing the annotation of the transcripts of a gene, a FASTA file containing the reference sequence, a FASTA file containing a set of RNA-seq reads, and four parameters L, K, ϵ and C (see Sect. 3), which are the minimum length of the considered MEMs, the parameter of consecutivity between two MEMs in the MEM-graph, the error threshold, and the minimum coverage, respectively. The tool computes the approximate alignments between the splicing graph and the RNA-seq reads, and outputs them in SAM format [11]. More precisely, a preprocessing module of the tool implemented in Phyton builds the splicing graph summarizing the annotated transcripts [7] from the GFF file and FASTA file of the reference sequence, and it also performs the transitive closure of the graph to represent all possible transcripts generated from the same set of exons. The new graph is the hypertext H to which the APMH algorithm is applied. The core part of SGAL implementing the algorithm consists of two steps: (i) the first one computes the MEMs between the reads and the serialization T_H of the hypertext and (ii) the second one performs the approximate pattern matching by processing the MEMs. As discussed in Sect. 3, to compute the MEMs between a text and a set of patterns, we used an external tool proposed by Ohlebusch *et al.* [14] and, therefore, we did not implement it. Since the second step is critical and requires to achieve good performance, we decided to implemented it in C++. Finally, a postprocessing module implemented in Python produces the output alignments in SAM format, in which one of the last fields is used to report the unannotated edges of the hypertext that are used by each alignment.

We performed an experimental analysis with the main goal of testing the use of SGAL to extract new splicing events from RNA-seq data w.r.t. a known annotation. For this purpose, we decided to apply the tool to real data consisting of

RNA-seq reads sequenced from the *Toxoplasma gondii* organism, whose annotation contains 8637 single-transcript genes[1]. The chosen dataset is part of a study which was designed to investigate the alternative splicing mechanism in this parasite, done by over-expressing a splicing factor protein, and then performing the high-throughput sequencing of the transcripts (RNA-seq) at different time points [20]. We selected one of the 3 replicates at the time 0 of this experiment, consisting of 15.4 million paired-end reads, having length 101bp (SRA accession id: SRR1407792). We selected genes covered by more than 5000 single-end reads, by aligning the RNA-seq sequences to the reference genome using STAR [6], a spliced aligner that can also run in an annotation-guided mode. To introduce more flexibility in the alignment process to select the set of interesting genes, we set a flanking region of 200bp in each considered gene, so that mappings exceeding the beginning or the end of the gene are considered. A total number of 938 genes resulted from this selection. Anyway, since this genomic portion will not be taken into account by SGAL, to perform the comparison with STAR, we run this latter tool giving as input the exact genomic region of each of the 938 selected genes, which resulted in a total of ~10 million of alignments by STAR (see Table 2). After that, SGAL has been applied to align each read P in the dataset to the related hypertext H, producing the best approximate pattern matchings to H having at most 7% of errors w.r.t. the length of the read P (since all reads have length equal to 100, this means identifying all matchings having edit distance smaller than 7).

Analyses were performed on a 64 bit Linux (Kernel 3.13.0) system equipped with Four 8-core Intel ® Xeon 2.30 GHz processors and 256 GB of RAM.

Table 1. Time and memory requirements for different combinations of the input parameters L (7, 10, and 15) and K ($\{8, 12, 17\}$, $\{11, 15, 20\}$, and $\{16, 20, 25\}$, respectively). Times are reported for both the computation of the MEMs and for the construction of the alignments, while for the memory it is shown the peak, which occurs during the alignment step.

L	K	MEM Time (s)	Aln. Time (s)	Aln. Memory (kB)	Num. Alns
7	8	1978	5371	43420	8706310
7	12	1978	5818	43831	8715527
7	17	1978	8156	50847	8718278
10	11	1690	1697	31491	8705929
10	15	1690	1724	31529	8715824
10	20	1690	1789	31768	8721981
15	16	1666	1600	31259	8703126
15	20	1666	1595	31265	8709622
15	25	1666	1612	31273	8714694

---

[1] Release 29 of ToxoDB annotation of TgondiiGT1.

We tested different combinations of the input parameters L and K to assess the time and space performance of our method on the aforementioned dataset of RNA-seq from the Toxoplasma organism. More precisely, we set L equal to 7, 10, and 15, and, for each of these values, we selected three values of K accordingly, that is $\{8, 12, 17\}$, $\{11, 15, 20\}$, and $\{16, 20, 25\}$, respectively. We also set the parameters ϵ and C equal to 0.07 and $(100 - 2 \cdot L)/100$, respectively.

For each tested combination of L and K, Table 1 reports times for the computation of the MEMs and the alignment steps, memory peak of the overall pipeline (which occurs in the alignment step), and the total number of alignments. As we can observe from Table 1 and, as it was expected, executions with lower values of L require higher time and memory, since an higher number of MEMs needs to be compared to find the best alignments. Moreover, given a fixed value of L, executions with higher values of K require higher time, since our algorithm tries to link MEMs considering a wider window of size K. Although this expected behavior is confirmed by almost all the runs, we note that the run with $L = 15$ and $K = 20$ does not abide by it. Nevertheless, this difference is marginal with respect to the whole time and does not depend on the implementation itself. Another conclusion that we can infer from Table 1 is that both L and K values do not have a huge impact on the number of alignments. In fact the minimum values have been obtained with $L = 15$, for which we observed that parameter K did not affect nor time (where the difference between minimum and maximum values is $\sim$15 s on a total of $\sim$1600 s) nor memory (where the values range from 31259 kB to 31273 kB), while the maximum time/memory consumption has been registered with $L = 7$ and $K = 17$ (8156 s time with 50847 kB of memory). Anyway, a good trade-off between time/memory and number of produced alignments can be achieved with $L = 10$ and $K = 20$.

Table 2. Results obtained by SGAL and STAR on the SRR1407792 dataset. The first two columns are parameters L and K used by SGAL, while the third and the fourth columns report the number of reads aligned by SGAL and STAR, respectively. The fifth column ("Common") corresponds to the number of common alignments between the two tools. The last five columns, namely "Clips", "Intron", "Mid-Intron", "Small-match", and "Mismatch", correspond to the alignments found by STAR and (as expected) not by SGAL.

L	K	SGAL	STAR	Common	Clips	Intron	Mid-Intron	Small-match	Mismatch
7	8	8872985	10040695	8709214	466325	546605	263604	2	54945
7	12	9023316	10040695	8720797	466277	546605	263495	2	43519
7	17	9324847	10040695	8726144	466266	546603	263318	2	38362
10	11	8810404	10040695	8705967	465721	546607	262274	927	59199
10	15	8872785	10040695	8716029	465700	546605	262136	864	49361
10	20	8950201	10040695	8722325	465 695	546605	262116	824	43130
15	16	8800848	10040695	8702965	465609	546676	262331	4089	59025
15	20	8836405	10040695	8709539	465598	546675	262311	3969	52603
15	25	8873383	10040695	8714715	465589	546675	262285	3894	47537

Although there is no other tool available that performs the same operations as SGAL, to evaluate its performance, we tried to compare the obtained results and also time and memory requirements with that of STAR, which solves a similar problem. For this reason a comparison with STAR in terms of accuracy of the results is not possible, due to the fact that they work on different reference sequences (genome in STAR versus transcripts in our method) and they also perform different operations (spliced alignment for STAR versus approximate pattern matching against a splicing graph for our method). However, to get a fair comparison of the two tools, we run STAR by giving as input the genomic region of each gene, the related annotation, and the related collections of reads that align to that region, as input data. To perform the spliced alignments on the selected genes STAR, with default parameters, required 7930 s with a peak of 262.5 MB of memory.

We also compared the quality of the results obtained by SGAL and STAR on the considered dataset. Notice that, since STAR aligns reads to a reference genome, it produces splice alignments that may involve all the genomic regions, such as exons as well as portions of introns, and it may use gaps even when aligning only to the exome. Moreover, since the main goal of SGAL is to compare reads with the annotation and understand how it is supported by the reads in terms of splice junctions or exon skipping events, at the moment we have not implemented specific criteria to map reads to the splicing graph. As a consequence of these aspects, in order to perform a fair investigation, we filtered the alignments of STAR, by considering only the primary ones, as well as the best one also for SGAL.

Table 2 reports the results obtained from the comparison between SGAL and STAR. For the aforementioned reasons, SGAL is not able to retrieve all the alignments obtained by STAR. We partitioned the primary alignments of STAR that are not found by SGAL: "Clips" (alignments having too long soft-clipping), "Intron" (alignments fully contained in intronic regions), "Mid-Intron" (alignments partially contained in intronic regions), "Small-Match" (alignments in exonic region but with small anchor regions), and "Mismatch" (alignments in exonic region having an high error rate). However, as it is possible to observe from the "Common" column, ∼99.99% of the SGAL alignments are found also by STAR. On the other hand, we were able to explain all the alignments of STAR (∼13%) that was not found by our tool, by assigning it to one of the previous categories.

As anticipated, from the output of SGAL, it is possible to detect the use of unannotated edges of the hypertext, added by the transitive closure. Such information can be extracted from the STAR output only through a rather complex post-processing step. For this reason, to have an even more fair comparison, we plan to extract potential novel splicing events from the STAR output. For such analysis, simulated input data will be generated, so that the quality results will be better assessed. Now, in our experimental analysis, since the considered splicing graphs consist of single transcripts, the unannotated edges can represent the skipping of one or more exons. More precisely, we considered only the read alignments on new edges for which the involved read has no other alignments that use only annotated edges.

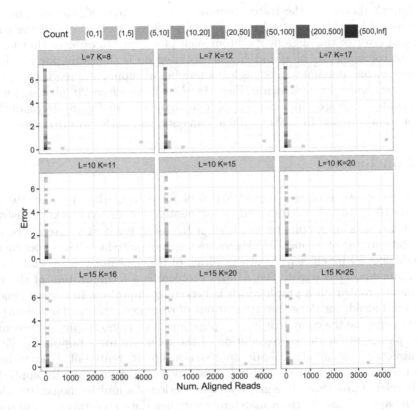

Fig. 3. Distribution of the unannotated edges found by SGAL for different combinations of L and K. Each plot reports the number of reads supporting the edge and the error of its alignment.

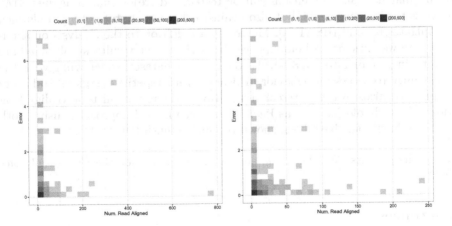

Fig. 4. Zoom of the plot in Fig. 3 with $L = 10$ and $K = 20$, limited to less than 1000 (left) and less than 300 (right) supporting reads.

Figure 3 shows, for the tested combinations of L and K, the unannotated edges found by SGAL, w.r.t. the number of reads supporting it and the corresponding error of the alignments. From such plots, we can observe that for the different combinations of L and K, the distribution is the same, for which the majority of the unannotated edges has a number of supporting reads lower than 1000. More precisely, as shown in Fig. 4 for $L = 10$ and $K = 20$, although many unannotated edge are supported by just one alignment, we found several of them with a good support, from 20 to 100 alignments, and with low error rate.

5   Conclusions

In this paper we proposed a practical tool for solving the approximate pattern matching to a hypertext and experimented it over hypertexts consisting of enriched splicing graphs to which align RNA-seq reads. The use of the succinct data structure to build MEMs makes the tool quite fast. The experimental analysis showed its efficiency both in terms of time/memory usage and also in finding unannotated edges. The APMH problem is a formalization of the read alignment problem to a graph which is becoming important in several genome analyses. Indeed, the linear representations of reference genomes fail (unsurprisingly) to capture the complexities of populations and, thus, graphs are becoming a new paradigm for the representation of the reference for a population [5]. In this direction, recent algorithmic approaches aim to represent genome information by indexing with an FM-index, which has been successfully applied to closely related problems [3], a graph representation of a multi-genome [16]. However the approximate pattern matching over such data structures is still under investigation. Due to the promising results obtained in the experimental tests we performed, we plan to extend SGAL to make it able to deal with the aforementioned categories, i.e. clips, totally or partially intronic reads, small matches, and mismatches. Such extension will be tested and (re)evaluated against STAR to assess the improvements, but also against other newer state-of-art tools such as TopHat2 [10] and HISAT [9]. Moreover, in addition to the analysis on simulated data we anticipated before, since the performance results we obtained are encouraging, we plan to run SGAL also on a human dataset, which will give us the opportunity to assess its behavior in dealing with repetitive regions and more complex scenarios. A future research direction is the use of our tool to align long reads from technologies such as PacBio on which hybrid approaches using both short and long reads have been shown to behave much better [15].

Acknowledgments. We thank the anonymous reviewers for their insightful comments.

References

1. Amir, A., Lewenstein, M., Lewenstein, N.: Pattern matching in hypertext. J. Algorithms **35**(1), 82–99 (2000)

2. Beretta, S., Bonizzoni, P., Della Vedova, G., Pirola, Y., Rizzi, R.: Modeling alternative splicing variants from RNA-seq data with isoform graphs. J. Comput. Biol. **21**(1), 16–40 (2014)
3. Bonizzoni, P., Della Vedova, G., Pirola, Y., Previtali, M., Rizzi, R.: LSG: an external-memory tool to compute string graphs for next-generation sequencing data assembly. J. Comput. Biol. **23**(3), 137–149 (2016)
4. Cormen, T.H., Leiserson, C.E., Rivest, R.L., Stein, C.: Introduction to Algorithms. MIT Press, 2nd edn. (2001)
5. Dilthey, A., Cox, C., Iqbal, Z., Nelson, M.R., McVean, G.: Improved genome inference in the MHC using a population reference graph. Nat. Genet. **47**(6), 682–688 (2015)
6. Dobin, A., Davis, C.A., Schlesinger, F., Drenkow, J., Zaleski, C., Jha, S., Batut, P., Chaisson, M., Gingeras, T.R.: STAR: ultrafast universal RNA-seq aligner. Bioinformatics **29**(1), 15–21 (2013)
7. Heber, S., Alekseyev, M., Sze, S.H., Tang, H., Pevzner, P.A.: Splicing graphs and EST assembly problem. Bioinformatics **18**(suppl. 1), S181–S188 (2002)
8. Horner, D.S., Pavesi, G., Castrignanò, T., De Meo, P.D., Liuni, S., Sammeth, M., Picardi, E., Pesole, G.: Bioinformatics approaches for genomics and post genomics applications of next-generation sequencing. Briefings Bioinf. **11**(2), 181–197 (2010)
9. Kim, D., Langmead, B., Salzberg, S.L.: HISAT: a fast spliced aligner with low memory requirements. Nat. Methods **12**(4), 357–360 (2015)
10. Kim, D., Pertea, G., Trapnell, C., Pimentel, H., Kelley, R., Salzberg, S.L.: TopHat2: accurate alignment of transcriptomes in the presence of insertions, deletions and gene fusions. Genome Biol. **14**(4), R36 (2013)
11. Li, H., Handsaker, B., Wysoker, A., Fennell, T., Ruan, J., Homer, N., Marth, G.T., Abecasis, G.R., Durbin, R.: The sequence alignment/map format and SAMtools. Bioinformatics **25**(16), 2078–2079 (2009)
12. Manber, U., Wu, S.: Approximate string matching with arbitrary costs for text and hypertext. In: Proceedings of the IAPR International Workshop on Structural and Syntactic Pattern Recognition, pp. 22–33 (1993)
13. Navarro, G.: Improved approximate pattern matching on hypertext. Theoret. Comput. Sci. **237**(1), 455–463 (2000)
14. Ohlebusch, E., Gog, S., Kügel, A.: Computing matching statistics and maximal exact matches on compressed full-text indexes. In: Chavez, E., Lonardi, S. (eds.) SPIRE 2010. LNCS, vol. 6393, pp. 347–358. Springer, Heidelberg (2010). doi:10.1007/978-3-642-16321-0_36
15. Rhoads, A., Au, K.F.: PacBio sequencing and its applications. Genomics Proteomics Bioinform. **13**(5), 278–289 (2015). sI: Metagenomics of Marine Environments
16. Sirén, J.: Indexing variation graphs. CoRR abs/1604.06605 (2016)
17. Thachuk, C.: Indexing hypertext. J. Discrete Algorithms **18**, 113–122 (2013)
18. Trapnell, C., Pachter, L., Salzberg, S.L.: TopHat: discovering splice junctions with RNA-seq. Bioinformatics **25**(9), 1105–1111 (2009)
19. Vyverman, M., De Baets, B., Fack, V., Dawyndt, P.: essaMEM: finding maximal exact matches using enhanced sparse suffix arrays. Bioinformatics **29**(6), 802–804 (2013)
20. Yeoh, L.M., Goodman, C.D., Hall, N.E., van Dooren, G.G., McFadden, G.I., Ralph, S.A.: A serine-arginine-rich (SR) splicing factor modulates alternative splicing of over a thousand genes in Toxoplasma gondii. Nucleic Acids Res. **43**(9), 4661–4675 (2015)

A Fast Algorithm for Large Common Connected Induced Subgraphs

Alessio Conte[1], Roberto Grossi[1], Andrea Marino[1(✉)], Lorenzo Tattini[2], and Luca Versari[3]

[1] Inria, Università di Pisa and Erable, Pisa, Italy
{conte,grossi,marino}@di.unipi.it
[2] IRCAN, CNRS UMR, 7284 Nice, France
lorenzo.tattini@unice.fr
[3] Scuola Normale Superiore, Pisa, Italy
luca.versari@sns.it

Abstract. We present a fast algorithm for finding large common subgraphs, which can be exploited for detecting structural and functional relationships between biological macromolecules. Many fast algorithms exist for finding a single maximum common subgraph. We show with an example that this gives limited information, motivating the less studied problem of finding many large common subgraphs covering different areas. As the latter is also hard, we give heuristics that improve performance by several orders of magnitude. As a case study, we validate our findings experimentally on protein graphs with thousands of atoms.

Keywords: Proteins · Structure similarity · Isomorphisms · Graphs · Listing

1 Introduction

Graph-based methods provide a natural complement to sequence-based methods in bioinformatics and protein modeling. Graph algorithms can identify compound similarity between small molecules, and structural relationships between biological macromolecules that are not spotted by sequence analysis [2]. These algorithms find motivation in the increasing amount of structured data arising from X-ray crystallography and nuclear magnetic resonance. Many examples of graphs fall under this scenario, such as chemical structure diagrams [4], 3D patterns for proteins [17], amino acid side-chains [1], and compound similarity for the prediction of gene transcript levels [25], to name a few.

Context for the Study. We are interested in common subgraphs between two given input graphs G and H. Recalling that a subgraph S of G is a subset of its nodes and connecting edges, S is said to be common with H if S is isomorphic to a subgraph of H: S is maximal if no other common subgraph strictly contains it, and maximum if it is the largest. The *maximum* common subgraph problem asks for the maximum ones, or simply for their size: this problem is classically

© Springer International Publishing AG 2017
D. Figueiredo et al. (Eds.): AlCoB 2017, LNBI 10252, pp. 62–74, 2017.
DOI: 10.1007/978-3-319-58163-7_4

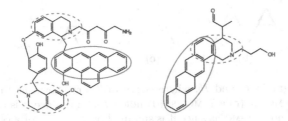

Fig. 1. Common structures in Liensinine derivatives. The two molecules share a tetracenic moiety (large circles) and an N-methylbenzopiperidinic moiety (smaller dotted circles). The maximum common structure is however a fairly common structure in organic molecules; the structure in the dotted circle which is maximal (but not maximum) is more likely to be interesting as it is more peculiar.

related to structural similarity. The *maximal* common subgraph (MCS) problem requires finding all the MCS's of G and H. The MCS problem can be constrained to *connected* and *induced* subgraphs (MCCIS) [8,16,17]: the latter means that all edges of G between nodes in the MCS are mapped to edges of H, and vice versa.

Maximum vs Maximal. Maximum common subgraphs are very often confused with MCS's, but they are *different* concepts: it is much faster in practice to find the maximum common subgraph size than all MCS's (e.g. [12]). As discussed later, while there are many results for the former, not much algorithmic research has been done in the past 20 years for MCS's, and the seminal results in [16,17] are still the state of the art. Note that a maximum common subgraph is not always meaningful as a structural motif, as it does not necessarily contain all the relevant or large common structures, as shown for the two molecules represented by the graphs in Fig. 1. In general, there may be arbitrarily large common substructures that give few information because of their frequent appearance in special type of macromolecules or polymers. Furthermore, when spotting structural motifs, it is not always possible to fix a priori the scoring system, and the maximum common subgraphs are not necessarily the ones getting the best score: a postprocessing can apply several scoring systems with a fast filtering and ranking of the MCCIS. This is more efficient than repeating a branch-and-bound search for each score.

Problem of Interest. It is well known that finding similar structures leads to highlighting similar biochemical properties and functions [22]. To this aim, we focus on *large* common connected induced subgraphs (LACCIS's). Indeed, considering *induced* and *connected* subgraphs reduces the search space and the number of solutions, while preserving the most significant ones [8,16,17]. To quickly find LACCIS's, we consider a modified version of the MCCIS problem: given a spanning tree T of G, we are interested in the common subgraphs between G and H for which their subgraph in G is connected using edges of T. We call these subgraphs T-MCCIS's (see Fig. 2 for an example).

For a set of spanning trees $T_1, \ldots, T_k$, we consider a set of LACCIS's such that each LACCIS L contains a T-MCCIS S for some $T \in T_1, \ldots, T_k$. In general, L satisfies $S \subseteq L \subseteq M$ for a MCCIS M, where $\subseteq$ denotes the containment relation

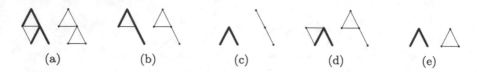

Fig. 2. (a) Two graphs G and H, where edges of T are shown as thicker; (b) a T-MCCIS that is also MCCIS; (c) a T-MCCIS; (d) not a T-MCCIS since it is a MCCIS but not spanned by T; (e) not a T-MCCIS since it is spanned by T but is not a common induced subgraph.

among induced subgraphs. Hence the larger L is, the closer is to a MCCIS. Note that L implicitly establishes an *isomorphism* between sets of matching nodes of G and H. However, this isomorphism is not unique: for example, if L is a clique of q nodes (i.e. all pairwise connected), it can be mapped through $q!$ isomorphisms.

When the nodes of G and H are labeled, the notion of LACCIS naturally extends by requiring, for instance, the nodes of the LACCIS to have matching labels, or considering a generalized compatibility function between nodes or edges.

Contributions. In this paper we provide an algorithm, called FLASH (Fast LAccis Searching Heuristic), that takes two connected labeled graphs G and H as input, along with some (random) spanning trees $T_1, \ldots, T_k$ of G, and returns a set of LACCIS's, where each LACCIS is represented as a pair of subsets of nodes, one from G and the other from H. For a spanning tree $T \in T_1, \ldots, T_k$, FLASH explores a variation of the product graph P [19] obtained from G and H, so that LACCIS are found as special cliques in P. FLASH does *not* materialize P, but navigates it implicitly to improve memory usage and running time, and produces T-MCCIS's at a fixed rate by employing a refined variation of an output-sensitive algorithm to find maximal cliques [11]. We remark that spanning trees have been previously employed to prune the search for frequent subgraphs [14], although such techniques do not extend to this problem.

FLASH applies the above approach to each tree, accumulating the found T-MCCIS's for $T = T_1, \ldots, T_k$. Then, it greatly reduces their number by a filtering criterion to make sense of the massive output: for a user-defined percentage σ (e.g. 70%), it selects a "covering" set of small size, such that each of the discarded T-MCCIS's has more than σ overlap with a retained one (priority is given to selecting larger sized ones). This filter shows that FLASH quickly finds solutions spanning different parts of G and H, whereas other approaches such as [17] tend to spend lot of time on the same nodes: small local additions and deletions of nodes produce a plethora of different subgraphs that significantly overlap.

Since FLASH could miss some maximal subgraphs (i.e. the ones not spanned by T_i for any i, as in Fig. 2d), it exploits the fact that the number of T-MCCIS's after filtering is relatively small, and performs a postprocessing to combine them with the purpose of enlarging the common subgraphs thus found (e.g. the subgraph in Fig. 2d is discovered if a non-spanned edge is spanned by another choice of T). The novelty of our approach is that the running time for a given spanning tree T is provably proportional to the number of reported T-MCCIS's, as confirmed by our experiments in Sect. 3. In other words, the more we pay in running time, the more T-MCCIS's we get. This is in contrast with the known algorithms for maximal common subgraphs that could be adapted to discover T-MCCIS's. They have the drawback of running into a computational black-hole, going through an explosive number of substructures even if there are few T-MCCIS's: for a T-MCCIS of k nodes, these algorithms have to potentially discard 2^k included subgraphs, and branch and bound does not help much in this case.

As a result, FLASH finds more solutions than other approaches and improves their performance by several orders of magnitude. When dealing with graphs of non-trivial size (e.g., thousands of nodes) we argue that the state-of-the-art approaches for maximal common subgraphs do not terminate within a conceivable time, thus making a practical comparison hard to perform. In our experiments, the size k of common subgraphs can easily be in the order of the hundreds (see Sect. 3) and FLASH performs well in practice even though its theoretical worst-case complexity is exponential.

Case Study with Proteins. The computational power of FLASH can bring benefits when modeling macromolecules such as proteins as graphs. To create a stress test for FLASH we adopted a fine-grained model, called all-atom, for representing the proteins 1ald, 1fcb, and 1gox from the Protein Data Bank (PDB), where the labeled nodes represent atoms within known secondary structures while the labeled edges represent covalent bonds (both backbone and non-backbone) as well as non-covalent interactions.

Current approaches benefit from a reduced computational load as they use coarse-grained models. For example, the 3D patterns of secondary structure elements in proteins have been modeled as graphs by using α-helices and β-strands, as nodes. These elements are approximately linear structures and they are represented as vectors in space, sometimes annotated with the length of their residues and hydrophobicity. As for the edges, they represent relationships between nodes expressed in terms of the angles and the distance between midpoints of the corresponding vectors [25]. In another representation, edges are calculated on the basis of contacts between the atoms in the respective structures/nodes, and indicate the spatial arrangements of the structures. In this way patterns can be also found in proteins with weaker similarities [17]. We refer the reader to Table 1 in [25] for a list of applications.

We think that exploring fine-grained models with FLASH, which was precluded with previous algorithms, can give finer details once data noise is filtered. However the design and validation of a fine-grained model is outside the scope of this paper, and deserves further independent study. Future investigations will be devoted also to the definition of a scoring function to rank the solutions produced by FLASH, and the application of our approach to problems where knowledge can be extracted from structural similarity, such as protein annotation.

Related Work. Both the problems of finding maximal or maximum common subgraphs has been studied for decades [5,9,25]. The corresponding decision version is NP-complete as it solves subgraph isomorphism problem, even on some restricted graph classes such as outerplanar graphs. The problem is difficult to approximate (MAX-SNP hard) even within a polynomial factor [15]. This motivates the search for effective heuristics.

Due to the strong connection between graph isomorphism and common substructures, the bioinformatics community has repeatedly expressed its interest in this problem from a computational point of view [7,12,15,17,21,25]. Finding the maximum common subgraph, as opposed to finding all maximal ones, allows for very effective cuts to the search space (e.g. branch-and-bound). This makes the computation much faster in practice, allowing researchers to process larger graphs with the available resources. However, as previously noted, their cut rules cannot be applied efficiently to LACCIS's.

Looking at previous work for graphs, it can be roughly classified into two categories: clique-based methods [17,23], non-clique-based backtracking methods [18,20,27].

Clique-based methods are widely employed and rely on the product graph P, transforming the common subgraphs of G and H into maximal cliques in P. This reduction dates back to the 70s [19] and has been shown to be effective on biological networks [12,17,23]. Algorithms such as the ones by Koch [16,17] are the state of the art [28], having laid down the basic principles for clique-based approaches; however, a tool able to efficiently enumerate LACCIS has not yet emerged. We will compare experimentally FLASH to the algorithms in [16,17]. For finding the maximal cliques, the algorithms by Bron and Kerbosch [6] or Carraghan and Pardalos [10] have been employed. We will use a refined variant of [11]. Cao et al. [9] observe that materializing P can be memory-wise expensive, and FLASH is able to avoid this issue.

Backtracking algorithms mostly build up on Ullman's strategy [24] for subgraph isomorphism (e.g. [20]). They often use branch-and-bound heuristics based on the specific requirements of the application at hand. The comparison in [12] shows how direct implicit methods for the maximum common subgraph, such as the one in [20], can outperform methods that exploit the product graph if the input graphs are small or contain many different labels. However, they do not apply efficiently to LACCIS's.

2   Methods

We give the main ideas behind FLASH for two labeled undirected graphs G and H, and a set $\{T_1, \ldots, T_k\}$ of (random) spanning trees of G. Let $T \in \{T_1, \ldots, T_k\}$.

Implicit Product Graph. We employ a variant of the transformation adopted by Koch [17] and borrowed from Levi [19], where we modify the color rule to take into account the edges of the spanning tree T. Define a *colored product graph* $P = G \cdot H$, with $P = (V_P, E_P)$. The nodes in V_P corresponds to ordered pairs of compatible nodes from G and H (e.g., with the same label), the first from G and the second from H, and the edges in E_P are as follows (where x and y denotes any two nodes in G and i and j any two nodes in H). There is a *black* edge in E_P between the nodes in V_P corresponding to (x, i) and (y, j) if

$\{x, y\} \in T$ (tree edge) and $\{i, j\} \in H$. There is a *white* edge in E_P between the nodes in V_P corresponding to (x, i) and (y, j) if either $\{x, y\} \in G \backslash T$ (non-tree edge) and $\{i, j\} \in E$, or both $x \neq y$ and $i \neq j$ are not connected by an edge in their graphs (resp. G and H). As in [17], there is a one-to-one correspondence between maximal cliques in P and maximal isomorphisms between subgraphs of G and H. The main difference is that in our case a maximal clique connected by black edges corresponds to a maximal subgraph connected by edges of T, instead of generic edges of G.

We call this kind of black-connected maximal clique a BC-*clique*, and reduce the problem of finding the T-MCCIS's to that of finding the isomorphisms/BC-cliques in the implicit P. We observe that the same T-MCCIS can give raise to several maximal isomorphisms/BC-cliques (matching the two sets of nodes in that T-MCCIS) that should be successively distilled to list it exactly once.

Building and navigating P is costly: P is a dense, massive graph with large maximum degree even when G and H are relatively small, sparse and with bounded degree. FLASH avoids storing P explicitly, and only stores G and H: it checks compatibility between assignments in constant time and iterates on neighbors in constant time per element by applying "on the fly" the rules used for generating P. This saves both memory and time, as G and H are much smaller and faster to access than P.

We now have to find the BC-cliques in $P = (V_P, E_P)$. Previous work on explicit product graphs employs a modified version of the Bron-Kerbosch algorithm [16] which does not perform *pivoting*, a pruning technique. The resulting algorithm thus iterates on every possible subset of each common subgraph, and its complexity and cost per solution are not clearly bounded. We do not reuse the above algorithm and use a different approach for FLASH as shown next.

Good Ordering. We use the spanning tree T to provide a *good ordering* of the nodes in $V(P)$. Let the nodes of G be numbered in such a way that, given a node u of G, there is only *one* edge of T between u and its neighbors u' with $u' < u$. This numbering can be computed by a pre-order visit of T, and induces the lexicographical order $\prec$ on the pairs (x, i) that corresponds to the nodes in $V(P)$. The *good ordering* of the nodes in V_P is obtained by numbering them consecutively in increasing lexicographical order $\prec$ of the corresponding pairs.

The T-based numbering of G's nodes induces an ordering of P's nodes that has the following *property*, whose proof is not given here for the sake of space. Let $P_{<v}$ denote the subgraph of P induced by the nodes $v' < v$, and $P_{<v} \cup \{v\}$ the one induced by $v' \leq v$, if C is a BC-clique in $P_{<v} \cup \{v\}$, then $C \backslash \{v\}$ is connected with black edges in $P_{<v}$.

This property ensures that every BC-clique can be found incrementally by FLASH: when adding a new node v to the set of BC-cliques found up to that point for the nodes of $P_{<v}$, two or more of the latter BC-cliques cannot be united because of the black edges incident to v, since the removal of v cannot disconnect the BC-clique. Hence we can consider just *one* of them to be extended by v, rather than any combinations of them. This speeds up significantly the computation.

Output-Sensitive Search. We perform an output-sensitive search of the BC-cliques by adapting to this problem a recent maximal clique enumeration algorithm [11]. The latter runs as fast as Bron-Kerbosch *with* pivoting, but it has the additional feature of guaranteeing that the total running time only depends on the number of returned BC-cliques's times a polynomial, meaning that a long execution will always yield many results. Furthermore, it allows us to define a parent-child relationship between partial solutions: this produces a stateless, memory-efficient search which avoids generating duplicate solutions. We adapted it to work on the implicit product graph P using the good ordering as it grows partial results by incrementally adding new nodes to them (see Sect. 4).

Filtering. It is important to distill all the T-MCCIS's found for each T. Those leading to the same LACCIS's are clearly redundant, and those that are small or mostly overlapping prevent us from making sense of a massive output. The FILTER procedure scans the found T-MCCIS's, giving priority to large ones and excluding the ones smaller than a given minimum size τ, and incrementally add them to a "cover" set if their overlap with every other isomorphism in the set is smaller than σ. Namely we retain T-MCCIS if, for either its subgraphs of G or H, the number of common nodes with any other T-MCCIS in the cover divided by its size is smaller than σ.

Recombining. As observed in the introduction, the T-MCCIS found may be fragments of larger (maximal) common subgraphs. To enlarge them, FLASH merges and uses FILTER on the output of all trees, then runs a RECOMBINE procedure which combines *compatible* T-MCCIS to generate larger LACCIS's. Two T-MCCIS's are compatible if they can be (partially) merged and their induced subgraphs in G and H are connected by one or more edges: RECOMBINE takes the largest part of the second T-MCCIS that can be added to the first and creates a larger LACCIS by merging them. This is repeated as long as new LACCIS's are created. After that, FILTER is applied again to remove redundant and partially overlapping isomorphisms, if any. We refer to the sequence of operations FILTER, RECOMBINE, FILTER as PROCESS. The final output of FLASH identifies LACCIS's composed of parts of the T-MCCIS's for $T = T_1, \ldots, T_k$.

3   Results

In this section, we describe our experimental results for FLASH. We considered the aggregated *raw* result after its output-sensitive search, and the *post*-PROCESS form after filtering and recombining the latter results by the method PROCESS (see Sect. 2). Specifically, we fixed $\tau = 10$ for the threshold on minimal size and $\sigma = 70\%$ for the overlapping threshold of FILTER (recall that PROCESS indicates the sequence FILTER, RECOMBINE, FILTER).

We chose to run FLASH with k random spanning trees for several values of k and with a set of spanning trees which forms a cover of the graph G (in our test case the number of spanning tree covering G was 5). We refer to the former variant as k-FLASH, with $k = 1, 3, 6, 12$, and to the latter as c-FLASH. As for the

Table 1. Experimental results

METHOD	RAW					post PROCESS				
	TIME (h)		MAX	H-IND	COUNT	TIME (h)		MAX	H-IND	COUNT
	PAR	WORK				PAR	WORK			
1ald vs 1fcb										
1-FLASH	0:12	0:13	63	59	9 215 182	0:01	0:06	70	39	17 468
3-FLASH	0:12	0:25	62	62	22 165 459	0:31	0:45	179	43	29 082
6-FLASH	0:29	1:01	59	58	47 329 927	0:24	1:05	155	48	41 080
12-FLASH	0:30	2:28	70	69	107 383 973	15:41	17:27	229	53	55 231
c-FLASH	0:13	0:40	55	54	38 315 376	0:09	0:43	118	42	41 410
KOCH	12	12	63	63	1 297 231	<0:01	<0:01	63	24	170
1ald vs 1gox										
6-FLASH	0:32	1:53	68	68	60 120 366	4:30	5:10	68	49	26 954
KOCH	12:00	12:00	64	64	4 775 963	<0:01	<0:01	64	6	6
1fcb vs 1gox										
6-FLASH	2:08	8:18	153	151	144 658 776	0:13	1:08	153	47	26657
KOCH	12:00	12:00	82	82	4 412 419	<0:01	<0:01	82	25	158
HelixD-1ald vs 1ald										
6-FLASH	2:01	9:00	171	170	58 925 057	0:07	0:25	171	38	6 561
KOCH	12:00	12:00	60	60	197 236	<0:01	<0:01	60	2	2
mod-HelixD-1ald vs 1ald										
6-FLASH	0:10	0:18	162	160	6 876 538	0:02	0:03	162	35	7 592
KOCH	12:00	12:00	60	60	81 884	<0:01	<0:01	65	2	2

raw result, FLASH runs k threads, one for each spanning tree, which (are forced to) terminate within a fixed time t, and then aggregates their results. We let each thread run for at most t/k hours, so that the bound for the overall CPU time is the same for all runs, with $t = 12$ h.

Following the discussion in the introduction, the baseline for the comparison of FLASH is Koch's algorithm [16], which produces LACCIS's using MCCIS and is the best algorithm known so far for MCCIS [28]. For a fair comparison, we optimized its implementation, denoted KOCH, so that it can use the implicit product graph P as well, noting that this optimization greatly improves its performance [26]; the computation is terminated after fixed time $t = 12$ h. Note that the RECOMBINE step has no effect on KOCH, whose output is made of MCCIS's that cannot be enlarged, and thus its additional time is negligible.

The above framework has been implemented in C++ and is available at github.com/veluca93/laccis. Our platform is a 24-core machine with Intel(R) Xeon(R) CPU E5-2620 v3 at 2.40 GHz, with 128 GB of shared memory. The system is Ubuntu 14.04.2 LTS, with Linux kernel 3.16.0-30.

We give a quick tour on our experimental study. First, we explain how we generated the data which has been used for the testbed of our experiments. Next, we show the experimental measures we have considered and we discuss how the choice of the spanning trees affects our analysis. We then analyze the experimental outcome for pairs of proteins, running FLASH and KOCH. Finally, we analyze the quality and consistency of the results returned with a protein and one of its sub-parts as input; also, we argue that FLASH is robust when perturbations of the input are introduced, e.g. node labels change.

Generating the Testbed Data. As mentioned in the introduction, we created a stress test with an all-atom fine-grained model for generating graphs from proteins from the PDB (www.rcsb.org). We thus exploited PDB data of 1ald, 1fcb (chain A), and 1gox proteins (which belong to TIM barrel families) to generate graphs where labeled nodes represent atoms within known secondary structures (as reported in PDB) while edges represent covalent bonds (both backbone and non-backbone) as well as non-covalent interactions.

We generated input graphs by means of **pdb2graph** [13]. First, PDB data is processed to generate edges from covalent bonds. Non-covalent interactions are estimated by extending the interaction distance up to 3.2 Å. Nodes are labeled with the element symbol and a secondary structure identifier. We thus generated 3 graphs with 2763 (9488), 3841 (12923), and 2696 (9059) nodes (edges) for 1ald, 1fcb, and 1gox respectively. Furthermore, we also considered two variants of a structure extracted from 1ald to test consistency and robustness of FLASH, discussed later.

Experimental Measures. For each pair of graphs in Table 1, we report the real execution time, that is the time (bounded by t/k hours for RAW) of the threaded execution PAR, and the total CPU time WORK (bounded by $t = 12$ h for RAW). Note that WORK of FLASH for RAW is less than t in all the cases as almost all the threads terminate earlier than the time limit t/k. We also report some analysis of the results, before and after applying PROCESS, in columns RAW and post PROCESS respectively. For each result set, we show the size of the greatest LACCIS in this set (i.e., MAX), the maximum h such that there are at least h LACCIS's of size h (i.e., H-IND), and the number of LACCIS's found (i.e., COUNT).

On the Choice of the Spanning Trees. Referring to the upper part of Table 1, given the pair of graphs 1ald and 1fcb, we compare the results of our k-FLASH for different values of k and of c-FLASH. It is worth observing (RAW column) that the number of T-MCCIS's found increases with the number of spanning trees used, recalling that c-FLASH uses 5 spanning trees, 6-FLASH and 12-FLASH produce a higher number of T-MCCIS's. After the post-processing, c-FLASH and 6-FLASH produce a similar number of LACCIS's, while 12-FLASH produces a larger number of LACCIS's but at the price of an higher post-processing time. For these reasons, we decided to focus on 6-FLASH in the remaining experiments in this section.

Running the Experiments. For the following pairs of graphs, 1ald vs 1fcb, 1ald vs 1gox, and 1fcb vs 1gox, we report the results for both KOCH and 6-FLASH. We remark how our algorithm, though heuristic, finds in this given time slot more LACCIS's than KOCH, whose result set is in theory complete. Moreover, it seems that FLASH is able to find larger LACCIS's than KOCH, as shown in the post PROCESS columns, where LACCIS's found by FLASH are greatly enlarged. Furthermore, it is clear that KOCH focuses the search on a limited portion of the graph, while FLASH is able to produce many more LACCIS's that do not overlap with each other (see COUNT in post PROCESS). For instance, consider the 1ald vs 1gox comparison. Even though KOCH finds 4 775 963 LACCIS's, after PROCESS we are left with just 6, of size at most 64: this means that all the remaining LACCIS's found by KOCH overlap with these 6 by at least $\sigma = 70\%$. This is not the case with FLASH, which obtains 26 954 LACCIS's after PROCESS. Even though our post PROCESS time is greater, this is compensated by the quantity and quality of our results, as well as the smaller running time of the algorithm.

Consistency and Robustness. To test the consistency and quality of the results, we extracted a portion of the protein 1ald (from Pro158 to Asn180), including an α-helix, and searched for it in the original protein. The corresponding subgraph has 171 nodes and 584 edges. Clearly, an effective algorithm must find at least a LACCIS which involves a large portion of the helix. Table 1 (HelixD-1ald vs 1ald) shows that our algorithm finds a LACCIS involving the whole helix (171 nodes), while this is not the case for KOCH in the time slot.

For the robustness, we introduced errors in the helix: we changed the labels of the alpha carbon atoms of Arg172 and Asn166 to a dummy label. We refer to this modified graph as mod-HelixD-1ald. A robust algorithm should not be significantly influenced by the introduced noise, and should find results similar to the ones obtained with HelixD-1ald. Table 1 (mod-HelixD-1ald vs 1ald) shows that our algorithm still finds almost the whole helix, while KOCH, although consistent with the previous result, finds just a small portion of the helix.

4   Implementation

In this section, we give more details about our algorithm to find BC-cliques in the implicit product graph P. First of all, we observe that a BC-clique in $P_{<v} \cup \{v\}$ is either a BC-clique of $P_{<v}$ or contains v. Moreover, a BC-clique C containing v is such that $C \backslash \{v\}$ is connected by black edges and so it is contained in a BC-clique of $P_{<v}$. Using the good ordering $v = v_1, \ldots v_p$ for the nodes of P, we can incrementally add nodes of P starting from an empty graph, and computing the new solutions using the previous ones. A simple implementation of this approach requires storing and scanning the whole set of BC-cliques found so far, which can be costly in practice. Note that we need the latter ones for two reasons: to avoid duplication of BC-cliques and to retrieve BC-cliques to be extended by the current node $v = v_i$. We give an alternative way as follows.

Let R_v denote the set of BC-cliques in $P_{<v} \cup \{v\}$. Each BC-clique $K' \in R_{v_i}$, for some i, is either a single node with no backward black edges or is generated from another BC-clique K in R_{v_j}, with $j < i$. Given a BC-clique K in R_{v_j}, we characterize which K' are generated from K. In particular, we identify for which $i > j$, K will generate a clique in R_{v_j}. We observe that a BC-clique K generates

---

Algorithm 1. Finding the BC-cliques in the implicit product graph P

Input: Two graphs G and H and a spanning tree T of G
Output: The BC-cliques in the implicit product graph $P = G \cdot H$ for T
$\pi \leftarrow$ lexicographic order of the nodes in V_P (good ordering);
for v_i *having no black edges going to* v_j *with* $j < i$ **do** visit($\{v_i\}$) ;

Procedure visit(K)
    if K *is maximal* **then** output K;
    Let N be the nodes larger than $\max(K)$ connected to K by a black edge;
    for $v_i \in N$ **do**
        Let C be the nodes in $K \cap N(v_i)$ that v_i can reach using black edges;
        if $C \cup \{v_i\}$ *is a real child of* K **then** visit $(C \cup \{v_i\})$;

---

a new BC-clique in $R_{v_i} \setminus R_{v_{i-1}}$ if there is a black edge from v_i to a node in K. Thus it is easy to find all nodes v_i that can be used to extend K using the black edges: they are the nodes that are connected to K with black edges and succeed the largest node in K in the good ordering. We call a BC-clique that is generated from K by adding one such node v_i a *potential child* of K: specifically, it is found by recursively extending $C \cup \{v_i\}$, where C is the set of nodes in $K \cap N(v_i)$ that v_i can reach using black edges. This allows us to generate the BC-cliques without using the sets R_{v_i} explicitly.

We now focus on how to avoid generating the same BC-clique twice. Given a BC-clique K', and letting v_i be the largest node of K' in the good ordering, we define the *parent* of K' as the BC-clique in $P_{<v_i}$ obtained from $K' \setminus \{v_i\}$ by recursively adding the smallest node that can fully extend the current (non-maximal) BC-clique. Note that the parent of a BC-clique K is unique and is a BC-clique in R_{v_j} with $j < i$. We avoid generating duplicates as follows: when trying to generate K' from K, we accept K' only if K is the parent of K'; otherwise, K' is discarded. As every BC-clique has exactly one parent, and can only be generated in one way from any BC-clique (when adding v_i), clearly it is impossible to generate any clique more than once. Hence we call a potential child K' of K a *real child* of K if it satisfies both conditions: K' is generated by adding v_i and cannot be extended with a node less then v_i, and the parent of K' is K. This technique to remove duplicates is similar to the reverse search approach used in several enumeration algorithms, first introduced in [3].

Our algorithm is a refined variant of [11] that takes into account the distinction between black and white edges and recursively examines all the BC-cliques as explained before. The roots of the recursion trees are given by all the nodes that have no black backwards edges. (The pseudocode in shown in Algorithm 1.) It is clear from what we said before that this algorithm generates exactly once all the BC-cliques of $P_{<v_i}$. When a (partial) BC-clique is generated, we check whether it is maximal in P, and if this is the case we output it; this way our algorithm outputs all BC-cliques of P in an output-sensitive fashion.

Acknowledgments. Work partially supported by projects MIUR PRIN 2012C 4E3KT (all authors except LT, LV) and UNIPI PRA_2015_0058 (authors RG, LT).

References

1. Artymiuk, P., Poirrette, A., Grindley, H., Rice, D., Willett, P.: A graph-theoretic approach to the identification of three-dimensional patterns of amino acid side-chains in protein structures. J Mol. Biol. **243**(2), 327–344 (1994)
2. Artymiuk, P., Spriggs, R., Willett, P.: Graph theoretic methods for the analysis of structural relationships in biological macromolecules. J. AM. Soc. Inf. Sci. Technol. **56**(5), 518–528 (2005)
3. Avis, D., Fukuda, K.: Reverse search for enumeration. Discrete Appl. Math. **65**(1), 21–46 (1996)
4. Bonchev, D.: Chemical Graph Theory: Introduction and Fundamentals. CRC Press, Boca Raton (1991)
5. Brint, A., Willett, P.: Algorithms for the identification of three-dimensional maximal common substructures. J. Chem. Inf. Comput. Sci. **27**(4), 152–158 (1987)
6. Bron, C., Kerbosch, J.: Finding all cliques of an undirected graph (algorithm 457). Commun. ACM **16**(9), 575–576 (1973)
7. Brun, L., Gaüzère, B., Fourey, S.: Relationships between graph edit distance and maximal common unlabeled subgraph. Technical report, HAL Id: hal-00714879, July 2012
8. Cao, Y., Charisi, A., Cheng, L., Jiang, T., Girke, T.: ChemmineR: a compound mining framework for R. Bioinformatics **24**(15), 1733–1734 (2008)
9. Cao, Y., Jiang, T., Girke, T.: A maximum common substructure-based algorithm for searching and predicting drug-like compounds. Bioinformatics **24**(13), i366–i374 (2008)
10. Carraghan, R., Pardalos, P.: An exact algorithm for the maximum clique problem. Oper. Res. Lett. **9**(6), 375–382 (1990)
11. Conte, A., Grossi, R., Marino, A., Versari, L.: Sublinear-space bounded-delay enumeration for massive network analytics: maximal cliques. In: ICALP (2016)
12. Conte, D., Foggia, P., Vento, M.: Challenging complexity of maximum common subgraph detection algorithms: a performance analysis of three algorithms on a wide database of graphs. J. Graph Algorithms Appl. **11**(1), 99–143 (2007)
13. Holder, L.: PDB-to-graph program (2015). https://github.com/mikeizbicki/datasets/tree/master/graph/pdb2graph. Accessed 04 May 2016
14. Huan, J., Wang, W., Prins, J., Yang, J.: Spin: mining maximal frequent subgraphs from graph databases. In: Proceedings of the tenth ACM SIGKDD International Conference on Knowledge Discovery and Data Mining, pp. 581–586. ACM (2004)
15. Kann, V.: On the approximability of the maximum common subgraph problem. In: Finkel, A., Jantzen, M. (eds.) STACS 1992. LNCS, vol. 577, pp. 375–388. Springer, Heidelberg (1992). doi:10.1007/3-540-55210-3_198
16. Koch, I.: Enumerating all connected maximal common subgraphs in two graphs. Theor. Comput. Sci. **250**(1), 1–30 (2001)
17. Koch, I., Lengauer, T., Wanke, E.: An algorithm for finding maximal common subtopologies in a set of protein structures. J. Comput. Biol. **3**(2), 289–306 (1996)
18. Krissinel, E., Henrick, K.: Common subgraph isomorphism detection by backtracking search. Softw.: Pract. Experience **34**(6), 591–607 (2004)
19. Levi, G.: A note on the derivation of maximal common subgraphs of two directed or undirected graphs. CALCOLO **9**(4), 341–352 (1973)

20. Mcgregor, J.: Backtrack search algorithm and the maximal common subgraph problem. Softw. Pract. Experience **12**, 23–34 (1982)
21. Raymond, J., Gardiner, E., Willett, P.: Rascal: calculation of graph similarity using maximum common edge subgraphs. Comput. J. **45**, 2002 (2002)
22. Sheridan, R., Kearsley, S.: Why do we need so many chemical similarity search methods? Drug Discov. Today **7**(17), 903–911 (2002)
23. Suters, W.H., Abu-Khzam, F.N., Zhang, Y., Symons, C.T., Samatova, N.F., Langston, M.A.: A new approach and faster exact methods for the maximum common subgraph problem. In: Wang, L. (ed.) COCOON 2005. LNCS, vol. 3595, pp. 717–727. Springer, Heidelberg (2005). doi:10.1007/11533719_73
24. Ullmann, J.: An algorithm for subgraph isomorphism. J. ACM **23**(1), 31–42 (1976)
25. Van Berlo, R., Winterbach, W., De Groot, M., Bender, A., Verheijen, P., Reinders, M., de Ridder, D.: Efficient calculation of compound similarity based on maximum common subgraphs and its application to prediction of gene transcript levels. Int. J. Bioinform. Res. Appl. **9**(4), 407–432 (2013)
26. Versari, L.: Ricerca veloce di pattern comuni a due grafi. Master's thesis, University of Pisa, Pisa, Bachelor Thesis (in Italian), University of Pisa (2015)
27. Wang, T., Zhou, J.: EMCSS: a new method for maximal common substructure search. J. Chem. Inf. Comput. Sci. **37**(5), 828–834 (1997)
28. Welling, R.: A performance analysis on maximal common subgraph algorithms. In: 15th Twente Student Conference on IT, University of Twente, The Netherlands (2011)

Ranking Vertices for Active Module Recovery Problem

Javlon E. Isomurodov[1,2], Alexander A. Loboda[1,2],
and Alexey A. Sergushichev[1,2(✉)]

[1] Computer Technologies Department, ITMO University, Saint Petersburg, Russia
{isomurodov,loboda,alserg}@rain.ifmo.ru
[2] JetBrains Research, Saint Petersburg, Russia

Abstract. Selecting a connected subnetwork enriched in individually important vertices is an approach commonly used in many areas of bioinformatics, including analysis of gene expression data, mutations, metabolomic profiles and others. It can be formulated as a recovery of an active module from which an experimental signal is generated. Commonly, methods for solving this problem result in a single subnetwork that is considered to be a good candidate. However, it is usually useful to consider not one but multiple candidate modules at different significance threshold levels. Therefore, in this paper we suggest to consider a problem of finding a vertex ranking instead of finding a single module. We also propose two algorithms for solving this problem: one that we consider to be optimal but computationally expensive for real-world networks and one that works close to the optimal in practice and is also able to work with big networks.

Keywords: Interaction networks · Active module · Vertex ranking · Dynamic programming · Integer linear programming · Connected subgraphs

1 Introduction

Network analysis has many applications in bioinformatics. This includes analysis of co-expression network for gene clustering [8], searching for reporter metabolites for metabolic processes [10], or stratification of tumor samples based on topological distance between somatic mutations in a gene interaction networks [5]. The overall idea is that by taking into account interactions between entities (genes, metabolites, etc.) one can better interpret the corresponding raw data (gene expression, metabolite concentrations, etc.).

One type of network analysis corresponds to the active module recovery problem. The goal of these methods is to find a connected subnetwork (module) that is enriched in individually important vertices. Such module, for example, could correspond to a signalling pathway for protein-protein interaction network [3] or a metabolic pathway for metabolic networks [7].

© Springer International Publishing AG 2017
D. Figueiredo et al. (Eds.): AlCoB 2017, LNBI 10252, pp. 75–84, 2017.
DOI: 10.1007/978-3-319-58163-7_5

There are many implementations for active module recovery [1,3,6,11]. These methods share a problem of non-monotonous dependence of the resulting module on the arbitrary significance threshold value. This means that when a method is rerun with a more relaxed threshold not only some vertices can appear, but they can disappear too. This situation is confusing for the user and makes interpretation of the results harder.

In this paper we consider a formulation of the active module problem in terms of connectivity-monotonous vertex ranking. This allows to generated modules for multiple thresholds that are consistent with each other. First, in Sect. 2.1 we formally define the problem and give related definitions. Then, in Sects. 2.2 and 2.3 we propose two methods to solve the problem: a brute-force-based method and semi-heuristic method based on solving a series of integer linear programming (ILP) problems. We also define two baseline methods in Sect. 2.4. Finally in Sect. 3 we compare the methods with each other and baseline methods on generated and real networks.

2 Methods

2.1 Formal Definitions

In this section we give a formal definition of the active module recovery problem in its ranking variant. Here we consider only networks with a simple structure of an undirected graph.

Let $G = (V, E)$ be a connected undirected graph and $w : V \to [0, 1]$ to be a weight function defined on its vertices. There is also an unknown connected subgraph (*active module*) and corresponding set of vertices M. Weights w are assumed to be random variables such that vertices from M are i.i.d. and follow a "signal" distribution and vertices from $V \setminus M$ are also i.i.d. but follow a "noise" distribution. Here we consider weights to be corresponding to P-values of a statistical test, where null hypothesis holds for vertices from $V \setminus M$ and corresponding weights follow uniform distribution $U(0, 1)$. Following [3] vertices from M are assumed to follow a beta-distribution $B(\alpha, 1)$ for some parameter α.

Definition 1. *Let $G = (V, E)$ be a graph. A* vertex ranking *of G is a permutation of its vertices V. For a ranking $r = (r_1, r_2, \ldots, r_{|V|})$ we consider vertices at the beginning of r (e.g. $r_1, r_2, \ldots$) to be more important and ranked higher than vertices at the end (e.g. $r_{|V|}, r_{|V|-1}, \ldots$).*

Definition 2. *Let us call a vertex ranking r of a connected graph G to be* connectivity-monotonous, *if all subgraphs G_k induced by vertices from ranking prefixes $r_{1..k} = (r_1, \ldots, r_k)$ for $k \in 1..|V|$ are connected.*

For convenience we will consider a rank prefix $r_{1..k}$ as a set $\{r_1, \ldots, r_k\}$ rather than a vector if the context requires it.

In this paper we will use AUC (Area Under the Curve) measure to define which ranking r of graph G better recovers module M.

Definition 3. *AUC value of a vertex ranking r for graph $G = (V, E)$ and module $M \subset V$ can be calculated using formula:*

$$AUC(r|M) = \sum_{i=1}^{n} \left(1 - \frac{|r_{1..i} \setminus M|}{|V \setminus M|}\right) \frac{[r_i \in M]}{|M|},$$

where $[r_i \in M]$ is equal to 1 if $r_i \in M$ and 0 otherwise.

To summarize we define the considered problem as follows.

Definition 4. *Given a connected graph G, an unknown active module M and vertices weights w that follow beta- and uniform distributions for vertices from M and $V \setminus M$ correspondingly, the ranking variant of the active module recovery problem consists in finding a connectivity-monotonous ranking r with the maximal value of $AUC(r|M)$.*

Later in this paper we consider the parameter α of the beta-distribution $B(\alpha, 1)$ to be known. Similarly to [3] one can infer parameters of the beta-uniform mixture from the vertex weights using maximum likelihood approach.

2.2   Optimal-on-Average Ranking

In this section we describe a method that finds ranking with the maximal expected value of AUC. Correspondingly, we call it *optimal-on-average* method.

First, let consider a set $D \subset 2^V$ of all vertex sets that induce a connected subgraph of G and a discrete probability $P(M)$ defined for all $M \in D$. Together this constitutes a probability space $\mathcal{M}$.

Our task is to find a ranking r with the maximal expected value of AUC score given a vector of vertex weights w:

$$E[AUC(r|\mathcal{M})] = \sum_{M \in D} P(M|w) \cdot AUC(r|M). \tag{1}$$

A conditional probability of a module $P(M|w)$ can be calculated using the Bayes' theorem:

$$P(M|w) = \frac{P(w|M) \cdot P(M)}{P(w)}$$

$$= \frac{P(M)}{P(w)} \cdot \prod_{v \in M} B(\alpha, 1)(w(v)) \cdot \prod_{v \in V \setminus M} U(0,1)(w(v)). \tag{2}$$

Let us rewrite the formula 1:

$$E[AUC(r|\mathcal{M})] = \sum_{M \in D} p(M|w) \sum_{i=1}^{n} \left(1 - \frac{|r_{1..i} \setminus M)|}{|V \setminus M|}\right) \frac{[r_i \in M]}{|M|}$$

$$= \sum_{i=1}^{n} \sum_{M \in D} \left(1 - \frac{|r_{1..i} \setminus M|}{|V \setminus M|}\right) \cdot \frac{p(M|w) \cdot [r_i \in M]}{|M|}. \tag{3}$$

This allows us to calculate $E[AUC(r|\mathcal{M})]$ iteratively:

$$E[AUC(r_{1..k}|\mathcal{M})] = E(AUC(r_{1..k-1}|\mathcal{M}))+$$

$$\sum_{M \in D|r_k \in M} \left(1 - \frac{|r_{1..k} \setminus M|}{|V \setminus M|}\right) \cdot \frac{p(M|w)}{|M|}. \quad (4)$$

Formula (4) allows to calculate every $r_{1..k}$ prefix ranking only one time.

This can be used to find the best ranking as shown in the Algorithm 1. There we fill in an array that for every set of vertices $D[i]$ from D contains a pair of values $dp[i].auc$ – expected AUC value of the best connectivity-monotonous ranking of vertices $D[i]$ and $dp[i].ranking$ – the corresponding ranking. The function $getArea()$ calculates the second summand of formula (4).

Algorithm 1. Optimal-on-average ranking.

```
1   procedure OptimalRanking(V,E):
2       D ← getConnectedSubgraphs(V, E)      ▷ elements of D ordered by size
3       dp[D] : (auc: Double, ranking: Vector)
4       for i = 1 to |D| do
5           M ← D[i]
6           forall v ∈ M do
7               if isNotConnected(M \ {v}) then
8                   continue
9               j ← get index of M \ {v} in D
10              auc ← dp[j].auc + getArea(D, dp[j].ranking, v)
11              if auc > dp[i].auc then
12                  v̄ ← (dp[j].ranking, v)
13                  dp[i] ← (v̄, auc)

14      return dp[|D|].ranking
```

The time complexity of the Algorithm 1 is $O(n^2 \cdot |D|^2)$. One call to $getArea()$ requires $O(n \cdot |D|)$ time and it is multiplied by $O(n \cdot |D|)$ for the outer loops.

2.3 Semi-heuristic Ranking

In this section we describe another approach for the vertex ranking problem. This approach is inspired by BioNet method [3] and consists in solving a series of integer linear programming (ILP) problems using IBM ILOG CPLEX library. Compared to the optimal-on-average approach from the previous section this method allows finding a ranking for large graphs in a rather reasonable time. As this method does not explicitly optimizes AUC score we call this method *semi-heuristic*.

First, similar to BioNet, let us find a subgraph of G that is most likely to be the active module. The most likely subgraph has the best (log)-likelihood score. The log-likelihood score of the module can be calculated as a sum of log-likelihood scores of the individual vertices in the module, where individual score for vertex v is calculated as:

$$score(v) = \log \mathcal{L}(\alpha, 1|w(v)) = \log(\alpha \cdot w(v)^{\alpha-1}).$$

Now, we can find a connected subgraph M with a maximal sum of vertex scores. This corresponds to an instance of Maximum-Weight Connected Subgraph problem (MWCS). This problem is NP-hard but it can be reduced to an ILP problem and solved by IBM ILOG CPLEX as, for example, in [4].

Using the found subgraph M we can define a crude partial ranking by saying that vertices of M go before $V \setminus M$.

Next, we define a procedure to refine such partial ranking. This procedure takes two sets of vertices: a set R that contains already ranked vertices and a set C that contain set of candidate vertices to be ranked. Then we find a subset X of C, so that $R \cup X$ is a connected and vertices from X should be ranked higher than $C \setminus X$.

Using this procedure we can recursively refine ranking up to the individual vertex level. Initially we solve an instance where R is set to an empty set and C contains all vertices. Then we do ranking for (R, X) and $(R \cup X, C \setminus X)$. We stop recursion when the candidate set consists of only one vertex.

A parameter of this procedure is how to select set X. For this end, similarly to the first step, we solve an MWCS instance, but with an additional constraint that requires the solution to contain at least one vertex from R and at least one but not all vertices from C. We set X as an intersection of the solution and the set C. The corresponding instance is solved by a modified solver from [9], where corresponding constraints were added into the ILP formulation.

Overall algorithm is shown as Algorithm 2. The procedure $findMaximum - SG()$ solves MWCS with the described additional constraints and returns chosen subset of vertices from C. If $list$ size is more than one, we call $refineRanking()$ to get a ranking of this set. The algorithm returns a ranking r of vertices C.

2.4 Baseline Methods

As base line for the experiments we consider the following two methods.

The first method ranks vertices by their weights: the smaller the weight, the higher is rank. This ranking is not connectivity-monotonous but is a good starting point. We will call this method non-monotonous.

The second method consists in running BioNet algorithm for ten different significance thresholds. As the BioNet modules $(M_1, M_2, \ldots, M_{10})$ can be non-monotonous we use the following combining procedure. We assign the highest rank to vertices from M_1, the second highest to M_2 M_1, the third to $M_3 \setminus (M_1 \cup M_2)$ and so on. The significance thresholds are selected to be distributed at equal steps between maximum and minimum log-likelihood vertex scores.

Algorithm 2. Semi-heuristic ranking refinement.

```
 1  procedure RefineRanking( V, E, R, C ):
 2  │   r : Ranking
 3  │   while C.size! = 0 do
 4  │   │   list ← findMaximumSG(V, E, R, C)
 5  │   │   if list.size > 1 then
 6  │   │   └   list ← refineRanking(V, E, R, list)
 7  │   │   r.addAll(list)
 8  │   │   R.addAll(list)
 9  │   │   C.removeAll(list)
10  │   return r
```

3   Experimental Results

We carried three series of experiments for different graph sizes. First, we considered small graphs of about 20 vertices where we were able to thoroughly compare all the considered methods. Next, we analyzed medium-sized graphs of 100 vertices. For such sizes that are closer to the real-world ones we analyzed all methods except optimal-on-average one, as it became computationally infeasible to run. Finally, we tested methods on a real-world graph of two thousand vertices.

3.1   Small Graphs

In the first experiment we have generated 32 different graphs of size 18. Then an active module of size 4 was chosen uniformly at random. Value of α was chosen from $U(0, 0.5)$ distribution. Vertex weights were generated from corresponding beta- and uniform distributions.

The results of the first experiment are shown on Fig. 1. They show that the optimal-on-average method in most cases works equal or better compared to both BioNet-like and non-monotonous baseline methods (top panels). The semi-heuristic method works similarly well compared to optimal (bottom-left panel) and better than BioNet-like method.

The distribution of active modules can be non-uniform in the real-world data, so we also carried out an experiment with such non-uniform distribution (see Sect. 3.4 for details). Aside from the four methods considered before we ran an optimal-on-average method parametrized by the real empirical distribution of the modules.

The results of this experiment are shown on Fig. 2. The situation is similar to the previous experiment with semi-heuristic method being close to optimal-on-average method and better than baseline methods. However, the semi-heuristic method works worse than optimal-on-average method parametrized by the real modules distribution.

3.2   Medium-Sized Graphs

Similarly to the previous section we have generated 32 different graphs of size 100. An active module were sampled to be the size of 5–25.

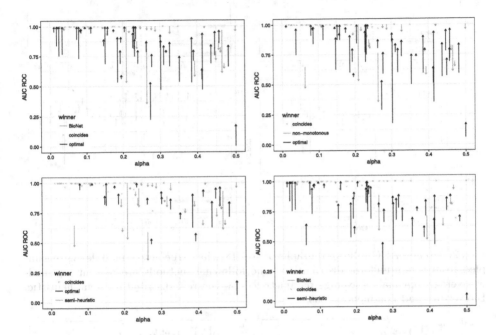

Fig. 1. Module AUC values for graphs of size 18. The following methods are present: optimal-on-average, semi-heuristic, BioNet-like and non-monotonous. Each panel shows comparison of two methods. One arrow correspond to one experiment with its ends corresponding to AUC value of the first and the second methods in the pair. The color depends on which method works better. True active modules were sampled from the uniform distribution.

On these graph sizes running the optimal-on-average method becomes infeasible, so we excluded it from the analysis. A median time of running the semi-heuristic method was 146 s.

The results of the experiment are shown on Fig. 3. Almost for all cases semi-heuristic ranking have worked better than both BioNet-like and non-monotonous baseline methods.

3.3   Large Real-World Graph

Finally, we analyzed performance of the proposed semi-heuristic method on the large real-word graph. For this experiment we used a protein-protein interaction graph from the example of BioNet package [2]. This graph has 2089 vertices and 7788 edges. An active module in this network was sample to be a size of 50–250.

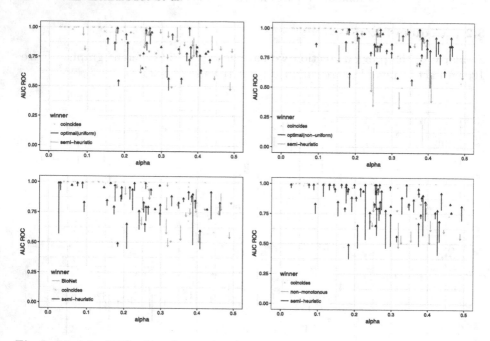

Fig. 2. Module AUC values for graphs of size 18 when true active modules were sampled from a non-uniform distribution. The following methods are present: optimal-on-average, optimal-on-average parametrized by the real distribution, semi-heuristic, BioNet-like and non-monotonous.

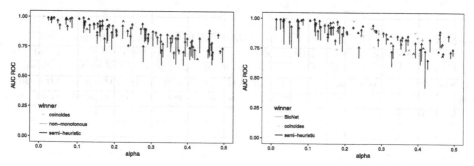

Fig. 3. Module AUC values for graphs of size 100. Three methods are present: semi-heuristic, BioNet-like and non-monotonous.

The results of the experiment are shown on Fig. 4. As for medium sizes semi-heuristic method works better than both baseline methods. On the other hand, the running time of the method increased significantly to about six hours.

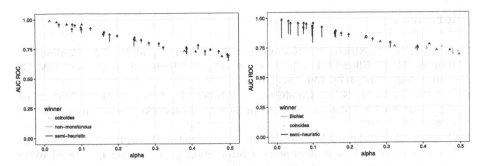

Fig. 4. Module AUC values for a real protein-protein interaction graph. Three methods are present: semi-heuristic, BioNet-like and non-monotonous.

3.4 Generating Graphs for Experiments

To mimic real network graphs generated for the experiments were scale-free. For the generation we used an existing implementation of the Barabasi-Albert algorithm from an R-package igraph.

For subgraph sampling of the given size we used the following procedure. Let $G = (V, E)$ be a connected graph, k be a required size of an active module and M is the set of vertices of the generated random active module. At the beginning M is empty. First we add into M a random vertex from the graph. Next we choose one of the adjacent vertex of M that does not already belong to M and add it. This step is repeated until M is of size k.

4 Conclusion

The problem of active module recovery appears in many areas of bioinformatics. Usually it is solved by an heuristic or exact algorithm that provides a module for a selected significance threshold. However, in practice multiple threshold values are tested and the results of these tests are not easily combined to be interpreted. In this paper we considered a ranking variant of this problem, where vertices are ranked before a particular threshold is selected. We also force a property of a module for a more stringent threshold to be a subgraph of a module for a less stringent one. We proposed two methods to solve this problem. The first method uses dynamic programming to find a ranking that maximizes an expected value of AUC score. We consider this method to be optimal, but it works only on small graphs. The second method does not explicitly maximize the AUC score but compares well to the optimal one and works better than the baseline methods in practice. However, it is also able to rank graphs with up to thousands vertices in a reasonable time.

Acknowledgements. This work was supported by the Ministry of Education and Science of the Russian Federation (agreement 2.3300.2017).

References

1. Alcaraz, N., Friedrich, T., Kotzing, T., Krohmer, A., Muller, J., Pauling, J., Baumbach, J.: Efficient key pathway mining: combining networks and OMICS data. Integr. Biol. **4**(7), 756–764 (2012)
2. Beisser, D., Klau, G.W., Dandekar, T., Muller, T., Dittrich, M.T.: BioNet: an R-package for the functional analysis of biological networks. Bioinformatics **26**(8), 1129–1130 (2010)
3. Dittrich, M.T., Klau, G.W., Rosenwald, A., Dandekar, T., Müller, T.: Identifying functional modules in protein-protein interaction networks: an integrated exact approach. Bioinformatics **24**(13), i223–i231 (2008)
4. El-Kebir, M., Klau, G.W.: Solving the maximum-weight connected subgraph problem to optimality (2014). arXiv:1409.5308
5. Hofree, M., Shen, J.P., Carter, H., Gross, A., Ideker, T.: Network-based stratification of tumor mutations. Nat. Methods **10**(11), 1108–1115 (2013)
6. Ideker, T., Ozier, O., Schwikowski, B., Siegel, A.F.: Discovering regulatory and signalling circuits in molecular interaction networks. Bioinformatics **18**(Suppl 1), S233–S240 (2002)
7. Jha, A.K., Huang, S.C., Sergushichev, A., Lampropoulou, V., Ivanova, Y., Loginicheva, E., Chmielewski, K., Stewart, K.M., Ashall, J., Everts, B., Pearce, E.J., Driggers, E.M., Artyomov, M.N.: Network integration of parallel metabolic and transcriptional data reveals metabolic modules that regulate macrophage polarization. Immunity **42**(3), 419–430 (2015)
8. Langfelder, P., Horvath, S.: WGCNA: an R package for weighted correlation network analysis. BMC Bioinformatics **9**(1), 559 (2008)
9. Loboda, A.A., Artyomov, M.N., Sergushichev, A.A.: Solving generalized maximum-weight connected subgraph problem for network enrichment analysis. In: Frith, M., Storm Pedersen, C.N. (eds.) WABI 2016. LNCS, vol. 9838, pp. 210–221. Springer, Cham (2016). doi:10.1007/978-3-319-43681-4_17
10. Patil, K.R., Nielsen, J.: Uncovering transcriptional regulation of metabolism by using metabolic network topology. Proc. Nat. Acad. Sci. **102**(8), 2685–2689 (2005)
11. Sergushichev, A.A., Loboda, A.A., Jha, A.K., Vincent, E.E., Driggers, E.M., Jones, R.G., Pearce, E.J., Artyomov, M.N.: GAM: a web-service for integrated transcriptional and metabolic network analysis. Nucleic Acids Res. **44**(W1), 194–200 (2016)

Computational Processes that Appear to Model Human Memory

John L. Pfaltz$^{(\boxtimes)}$

Department of Computer Science, University of Virginia, Charlottesville, USA
jlp@cs.virginia.edu

Abstract. This paper presents two computable functions, ω and ε, that map networks into networks. If all cognition occurs as an active neural network, then it is thought that ω models long-term memory consolidation and ε models memory recall. A derived, intermediate network form, consisting of chordless cycles, could be the structural substrate of long-term memory; just as the double helix is the necessary substrate for genomic memory.

Keywords: Closure · Chordless cycles · Long-term memory · Consolidation · Recall · Neural network

1 Introduction

There seems to be consensus that our sensations, ideas, and memories are really just active networks of neurons in our brains [12,30]. And we have a good idea where in the brain specific kinds of mental activity occur, *e.g.* [17,28] But, to our knowledge, no one has any idea as to what kinds of networks correspond to any specific sensation, concept or memory.

We know that neurons can stimulate other neurons by means of electric (or chemical) charges proceding along an axon to one, or more, synapses [29]. That would suggest that a directed, asymmetric network is a reasonable model. However, such an asymmetric network may best model neuronal behavior, but not neuronal state. Many neurons are interconnected by dendrites. These are thought to be bi-directional, thus implementing symmetric relationships that may recognize a state necessary to activate a neuron.

Given this state of uncertainty, we have chosen to explore symmetric relationships, or graphs or networks, in this paper. Some of the mathematical results we present may be true as well for asymmetric (directed) networks; some would require minor rewording; and some will no longer be true at all.

Regardless of whether our neural networks are essentially symmetric or asymmetric, it would appear that a mathematical treatment of networks, or graphs, or relationships is a fruitful way to approach them. That we will do in this paper.

In Sect. 2, we clarify our interpretation of relationships and their visual representation as graphs or networks. We also introduce the concept of "closure". In Sect. 3, we describe a computational process, ω, which reduces any network

D. Figueiredo et al. (Eds.): AlCoB 2017, LNBI 10252, pp. 85–99, 2017.
DOI: 10.1007/978-3-319-58163-7_6

to its unique, irreducible "trace". We will claim that this procedure appears to model the process of long-term memory "consolidation".

The ω process is a well-defined function over the space of all finite networks in that for any network $\mathcal{N}$, ω yields a unique irreducible trace $\mathcal{T}$. Thus the inverse set, ω^{-1}, defines the abstract set of all networks that reduce to the same specific irreducible trace. In Sect. 4, we present a computational process which generates specific members within ω^{-1}. We will argue that this can model "memory reconstruction".

In Sect. 5 we present additional evidence to support our claims to model long-term memory consolidation and recall. Certain mathematical details are developed in an appendix, Sect. A.

2 Sets, Relations, and Closure

Our computations are set based. The nature of the elements comprising the sets play no part, and can be quite arbitrary. So unlike most computational systems in which the variables will be `int` or `float`, our variables have type `setid`. We program using a set manipulation package in C++ with operators such as `is_contained_in` and `union_of`. Sets themselves are represented as extensible bit strings, so that the operators above are effectively of order O(1). There is no theoretical upper bound of these sets, but we have not tested it with sets of cardinality exceeding 50,000. A somewhat fuller description is given in [19]. All the following set-based operators and procedures have been implemented, and fully tested, using this system.

We use a standard set notation. A **set** S is comprised of elements $\{a, b, \ldots, y, z\}$ of unspecified type. The curly braces { } indicate that these elements are regarded as a "set". Sets are denoted by upper case letters, e.g. X, Y; elements are always lower case, e.g. x, y. Sometimes we elide the commas, as in $Y = \{abc\}$.

If an element x is a **member of** the set X, we write $x \in X$. If a set X is **contained in** another set Y, that is, $x \in X$ implies $x \in Y$ (here x is a variable running over all elements of X), we write $X \subseteq Y$. If the containment is **strict**, that is there exists $y \in Y$, $y \notin X$, we write $X \subset Y$. By $X \cup Y$ and $X \cap Y$ we mean the union and intersection (meet) of X and Y respectively.

One may have a "set of sets", which we call a **collection**, and denote with a caligraphic letter. Thus we may have $X \in \mathcal{C}$.

2.1 Relationships

Let S be any set. A **relation**, η, on S is a function, which given any subset $Y = \{y_1, y_2, \ldots, y_k\} \subseteq S$ returns the related set $Y.\eta = \{z_1, z_2, \ldots, z_n\} \subseteq S$. This is a bit unusual. It is more common to think of relations as links, or edges, between elements, such as illustrated by the undirected graph, or network, of Fig. 1, which we will use as a running example. η is sometimes regarded as a set of "edges" in a graph theoretic approach. But we prefer to define relations in terms of sets and functional operators. It provides an additional measure of

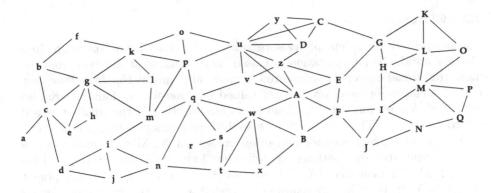

Fig. 1. A very small relationship η, or network $\mathcal{N}$, on 43 nodes, or elements

generality which can be of value. We emphasize this set-based definition by using suffix notation, such as $Y.\eta$ to mean the set of elements $\{z\}$ that are related to Y by η. We call $Y.\eta$ the **neighborhood** of Y. In Fig. 1, $\{g\}.\eta = \{bcefghklm\}$ and $\{IM\}.\eta = \{FHIJLMNOPQ\}$.

If η has the property that

$$\text{P1: } Y.\eta = \bigcup_{y \in Y}\{y\}.\eta \qquad \text{(extensibility)}$$

that is, $Y.\eta$ is the union of all the subsets $\{y\}.\eta$ for all $y \in Y$, we say that η is **extensible**, or *graphically representable*, so that Fig. 1 is an accurate representation of η.

If the relationship is not extensible, then it constitutes a "hypergraph" [3,7]. To more easily illustrate the concepts of this paper with graphs, we will assume that η is extensible; but unless explicitly noted none of the mathematical assertions require it. Moreover, we observe that for large, sparse relationships, matrix representations and operations are quite impractical [34].

In addition to extensibility, P1, a relationship η may also have any of the following 3 properties: that for all $X, Y \subseteq S$,

P2: $Y \subseteq Y.\eta$ (expansive or reflexive)[1]
P3: $X \subseteq Y$ implies $X.\eta \subseteq Y.\eta$ (monotone)[2]
P4: $X.\eta = Y$ implies $Y.\eta = X$ (symmetric)[3]

The relation of Fig. 1 is symmetric; its graph is **undirected**. By a **network**, $\mathcal{N} = (N, \eta)$, we mean a set N of nodes or elements, together with any relationship η. For this paper, we require that η satisfies the functional properties P2, P3 and P4.

---

[1] This is primarily for mathematical convenience.
[2] Probably essential. If η is not monotone, we can prove very few mathematical results of interest.
[3] Unnecessary, relaxed in other papers such as [25,26].

2.2   Closure

The mathematical concept of "closure" plays a key role in our approach. In a discrete world, the interpretation of closed sets is somewhat different from the more traditional concepts encountered in classical point-set topology. Our view is that a closure operator, φ, is a set-valued function whose domains are also sets. If Y denotes any set, $Y.\varphi$ denotes its closure; that is the smallest closed set containing Y. Thus, like η, φ is a well-defined function mapping subsets, $X, Y \subseteq N$ of a given network into other subsets of N. More formally, φ is a **closure operator** that satisfies the following 3 closure axioms, C1: expansive ($Y \subseteq Y.\varphi$), C2: monotone ($X \subseteq Y$ implies $X.\varphi \subseteq Y.\varphi$), and C3: idempotent ($Y.\varphi.\varphi = Y.\varphi$). Readily, any relationship operator, η, satisfying properties P2 and P3 is *almost* a closure operator. It has only to satisfy the idempotency axiom. But normally, $Y.\eta \subset Y.\eta.\eta$ since neighborhoods tend to grow.

An alternative definition of closure asserts that a collection $\mathcal{C} = \{C_1, \ldots, C_n\}$ can be regarded as the closed sets of a superset S if and only if C4: the intersection $C_i \cap C_k$ of any these closed sets is itself closed (in $\mathcal{C}$). It is not difficult to prove that C4 implies C1, C2, and C3, and conversely.

We normally think of closure in terms of its operator definition. Because φ is expansive, C1, the superset S must be closed; by C4, if any two closed sets are disjoint, the empty set $\emptyset$ must also be closed.

2.3   Neighborhood Closure

One important closure operator, φ, called **neighborhood closure** can be defined with respect to network relationships. We let

$$Y.\varphi = \bigcup_{z \in Y.\eta} \{\{z\}.\eta \subseteq Y.\eta\} \tag{1}$$

That is, if $z \in Y.\varphi$ then z is not related to any elements that are not already related to Y. Convince yourself that in Fig. 1, $\{c\}.\varphi = \{ac\}$, $\{g\}.\varphi = \{eghl\} \subset \{g\}.\eta$ and $\{u\}.\varphi = \{uyD\}$. It is not hard to show that φ, so defined with respect to η satisfies the closure axioms C1, C2 and C3 and that for all $\{y\}$, $\{y\} \subseteq \{y\}.\varphi \subseteq \{y\}.\eta$.

3   Irreducible Networks

If a singleton set $\{y\}$ is not closed, say $z \in \{y\}.\varphi$, then $\{z\}.\varphi \subseteq \{y\}.\varphi$, so z contributes little to understanding the structure of η in terms of closure. In Fig. 1, $\{a\}.\varphi = \{a\} \subset \{ac\} = \{c\}.\varphi$. Removing a, and its connections, results in minimal information loss with respect to η as a whole.

We say a network $\mathcal{N} = (N, \eta)$ is **irreducible** if every singleton set, $\{y\}$, is closed. That is, if for all $y \in N$, $\{y\}.\varphi = \{y\}$. In Fig. 1, $\{f\}.\varphi = \{f\}$, but from observations above, the entire network is not irreducible.

If $\{y\}$ is not closed, only elements z in $\{y\}.\eta$ could possibly be in $\{y\}.\varphi$ so only those need be considered. If $\{z\}.\eta \subseteq \{y\}.\eta$ so that $\{z\}.\varphi \subseteq \{y\}.\varphi$, we say z is **subsumed** by y, or z **belongs** to y. We can remove z from N, together with all its connections, and add z to $\{y\}.\beta$, the set of all nodes belonging to $\{y\}$ which we call its β-**set**. Since, $y \in \{y\}.\beta$, its cardinality, or β-**count**, $|\{y\}.\beta| \geq 1$, a value we will use in the next section. We use the pseudocode of Fig. 2 to implement the process ω that reduces any network $\mathcal{N}$ to its irreducible core, which is called its **trace**, $\mathcal{T}$.[4] This version of ω only records β-counts, not entire β-sets.

```
while there exist reduceable nodes
    {
    for_each {y} in N
        {
        get {y}.nbhd;
        for_each {z} in {y}.nbhd - {y}
            {
            if ({z}.nbhd contained_in {y}.nbhd
                {        // z is subsumed by y
                remove z from network;
                |{y}.beta| = |{y}.beta| + |{z}.beta|;
                }
            }
        }
    }
```

Fig. 2. Reduction code, implementing ω

The irreducible trace, $\mathcal{T}$, of Fig. 1 is shown in Fig. 3. The trace is the network on 26 elements with bolder connections. β-counts ≥ 2 are shown in square brackets, [ ]. β-sets are delineated with dashed lines. Observe that $\{g\}$ has subsumed e, h and l, so that $|\{g\}.\beta| = |\{eghl\}| = 4$ while $\{p\}$ has subsumed o, so $|\{p\}.\beta| = |\{op\}| = 2$. A total of 17 nodes were subsumed and eliminated.

The order in which individual nodes y are examined is arbitrary. Thus, one can create networks that require $n = |N|$ iterations of the outer loop. So this process has a theoretical complexity of $O(n^2)$. However, in tests with rather complex networks of several thousand nodes, the maximal number of iterations has never exceeded 7. Its effective complexity appears to be quite reasonable. Moreover, because of its local nature, the inner loop could be easily implemented in parallel.

We keep speaking of the *function* ω. It can be shown (see Sect. A), that for any network $\mathcal{N}$, its irreducible trace, $\mathcal{T}$, is *unique* (up to isomorphism). Therefore, the pseudocode of Fig. 2 does indeed embody a well-defined computational function which we denote by ω. Not only is ω a function, we can actually characterize its

---

[4] In [24], this was called the "spine" of $\mathcal{N}$.

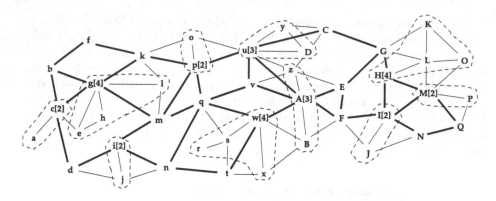

Fig. 3. The irreducible trace $\mathcal{T}$ of Fig. 1

output, $\mathcal{T} = \mathcal{N}.\omega$. When η is symmetric, if y is a node in $\mathcal{T} = \mathcal{N}.\omega$, then y will be either: (a) an isolated node; (b) an element of a chordless cycle of length ≥ 4; or (c) an element in a path between two chordless cycles of length ≥ 4 (again see Sect. A).

A **chordless cycle** is most easily visualized as a necklace of pearls (or beads). More formally, it is a sequence $< y_1, y_2, \ldots, y_n, y_1 >$ where $y_{i\pm 1} \in \{y_i\}.\eta$, $y_1 \in \{y_n\}.\eta$ and $y_{i\pm k} \notin \{y_i\}.\eta$ if $k > 1$. In Figs. 1 and 3 the sequence $< c, d, i, m, g, c >$ is a chordless cycle of length 5 and $< b, c, d, i, m, p, k, f, b >$ is one of length 8. Granovetter [13] called chordless cycles the "weak connections" of a social network. He felt they were the key to understanding the network structure as a whole. Chordless cycle structures have been relatively unstudied, while "chordal graphs" (with no chordless cycles) have an exhaustive literature [16]. Even when η is not symmetric, chordless cycles are basic to the characterization of an irreducible trace [26].

This trace, $\mathcal{T}$, of chordless cycles preserves a number of important properties found in the original network, $\mathcal{N}$. First, it preserves the shortest path structure between retained nodes. Consequently, connectivity and the distances between nodes (as usually defined) are preserved. Further, "network centers", [2,8,9], whether with respect to distance or "betweenness", are preserved in the trace.

The physical nature of human long-term memory is not at all a settled matter. We are fairly certain that the hippocampus of the brain is heavily involved [11,28]; but just how is rather unclear. One school of thought posits that long-term memories are recorded in some form of "memory trace" [4,32,36]. But, because no trace of these supposed "memory traces" has ever been physically detected (pun intended), others disbelieve this theory [5,18].

There is more consensus that some form of processing which distinguishes long-term memory from short-term memory does occur. This process is commonly called **consolidation** [1,14,18]. We believe that ω is analogous to consolidation, and that chordless cycles, in some form, are analogous to the elusive "memory trace", whence our terminology.

4   Computing Similar Networks

Since ω is a well-defined function mapping the space of all finite, symmetric networks into itself, one can consider $\mathcal{N}.\omega^{-1}$, which is the collection of all networks $\mathcal{N}_i$ such that $\mathcal{N}_i.\omega = \mathcal{T} = \mathcal{N}.\omega$. Two such networks, $\mathcal{N}_i$ and $\mathcal{N}_k$, that have the same irreducible trace are said to be **structurally similar**. Readily, structural similarity is an equivalence relation. Even though $\mathcal{N}_k$ may be similar to $\mathcal{N}_i$, they may have very different cardinalities. A network $\mathcal{N}_k(N_k, \eta_k)$ is said to be **strongly similar** to $\mathcal{N}_i(N_i, \eta_i)$ if $\mathcal{N}_k.\omega = \mathcal{N}_i.\omega$ and $|N_k| = |N_i|$.

```
for all {y} in N
{
    while (|{y}.beta| > 1)
    {
        create new node z;
        S = choose_random_in ({y}.nbhd);
        {z}.nbhd = S;
        k = random_int(1, |{y}.beta|-1);
        |{y}.beta| = |{y}.beta| - k;
        |{z}.beta| = k;
        add {z} to N;
    }
}
```

Fig. 4. Pseudocode for ε which generates strongly similar networks.

The pseudocode above in Fig. 4 describes a computational process ε that, given the trace $\mathcal{T}$ of a network $\mathcal{N}$ together with β-counts, randomly expands it to a strongly similar network $\mathcal{N}' = \mathcal{N}.\omega.\varepsilon$. The process choose_random_in returns a random subset of its argument. Since $\{z\}.\eta = S \subseteq \{y\}.\eta$, the node z will be subsumed by (or belong to) y if reduced again ensuring that $\mathcal{N}.\omega.\varepsilon.\omega = \mathcal{N}.\omega$. When a node $\{y\}$ is expanded, its β-count is decremented, and if > 1, part of the remainder may be added to the β-count of $\{z\}$. Consequently, by creating just as many new nodes as had belonged to any node $\{y\}$, we ensure that $|N'| = |N|$. This kind of ε process has been called an "expansion grammar" in [22]. The construction of ε, where $\{z\}.\eta \subseteq \{y\}.\eta$, assures us that $\mathcal{T}.\varepsilon.\omega$ will be $\mathcal{T}$ again. Consequently, for any network $\mathcal{N}' = \mathcal{T}.\varepsilon$, $\mathcal{N}' \in \mathcal{N}.\omega^{-1}$, so $\mathcal{N}'$ and $\mathcal{N}$ are structurally similar.

Let $\mathcal{N}$ be the network of Fig. 1. The following Fig. 5 shows a network $\mathcal{N}'$ that was randomly expanded by ε, given the irreducible trace $\mathcal{T}$ of Fig. 3.

The numbered nodes were appended to the trace and roughly correspond to the 17 subsumed nodes. $\mathcal{N}'$ is strongly similar to the network $\mathcal{N}$ of Fig. 1 because $\mathcal{N}'.\omega = \mathcal{T} = \mathcal{N}.\omega$.

Such a semi-random "retrieval" process may be inappropriate in computer applications [26], but it seems to model biological recall rather well. It has been

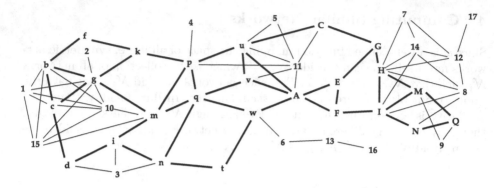

Fig. 5. A reconstructed network $\mathcal{N}' = \mathcal{T}.\varepsilon$ that is strongly similar to $\mathcal{N}$ of Fig. 1

observed that the recall and reconstruction of our long-term memories is seldom exact [14]. Our memories often are confused with respect to detail, even when they are generally correct. Reconstruction of a network trace by ε has these very properties.

Given that for all networks $\mathcal{N}$, $\mathcal{N}.\omega.\varepsilon.\omega = \mathcal{N}.\omega = \mathcal{T}$, it also supports the notion of "re-consolidation" which asserts than long-term memories are repeatedly recalled and re-written with no change, unless deliberately distorted in our (semi)conscious mind [18, 35].

5   Biological Memory

A computational model need not actually explain the behavior that it models. For example, the path of a thrown projectile has an excellent parabolic model. However, further study of this conic formulation contributes little to the understanding of either gravity or air resistence. By the same token, there need not be closure operators or chordless cycles involved in the performance of human memory, for the model to be valid. But, it would be a powerful verification of this model if we could demonstrate the existence of chordless cycle structures in a memory representation. We can't. Neither, to our knowledge, does anyone else know the structural format of our long-term memory.

Throughout this paper we have suggested parallels found in various memory studies. But, do these computational processes, ω and ε, really model biological memory? We just don't know. In this section we offer a few more tantalizing clues.

5.1   Role of Closure

We employed "closure" as the basic mathematical concept in the preceeding development. But, are instances of closure actually found in biological organisms? We offer two suggestive examples.

First, The visual pathway consists of layers of cells, beginning with the rods and cones of the retina passing stimuli toward the primary visual cortex. The neurological structure of this visual pathway is reasonably well understood, *c.f.* [12, 29, 33]. The individual functions of its layers are less well so.

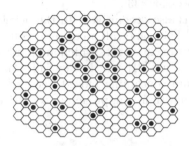

Fig. 6. Excited cells in a cross section of the visual cortex.

Imagine that Fig. 6 depicts a cross section of the retinal region. Dark cells denote visually excited cells. Although tightly packed, the actual neuronal structure is not as regular as this hexagonal grid; but this regularity plays no part in the process.

Let α be an existential operator defined as $Y.\alpha = black$ (excited), if and only if $\exists z \in Y.\eta$ where z is *black* (excited). Let β be the existential operator defined by $Y.\beta = white$ (quiescent), if and only if $\exists z \in Y.\eta$ such that z is *white* (quiescent). Figure 7(a) illustrates the excited (small ×) neighbors of Fig. 6. Figure 7(b) illustrates $Y.\alpha.\beta$ in which all excited cells of Fig. 7(a) that have at least one quiescent (white) neighbor become quiescent (white). The resulting central figure becomes evident; it is a closed object, because the pair of operators $(\alpha.\beta)$ is a closure operator. The pair $(\alpha.\beta)$ is idempotent because iterating them, as in $Y.(\alpha.\beta).(\alpha.\beta)$ yields no new black (excited) cells.

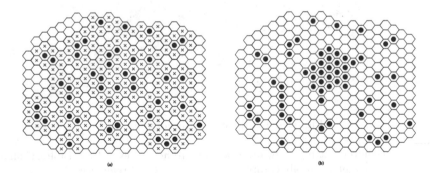

Fig. 7. (a) $Y.\alpha$, excited cells, (b) $Y.\alpha.\beta$, remaining excited cells.

This two step operation can occur at the neural firing rate. It is an effective parallel process that was first proposed to eliminate salt and pepper noise in computer imagery [31]. Readily, such a "blob detection" capability would have evolutionary value. Does such a capacity exist? We don't know for sure. But, it is thought that the visual pathway is organized in a manner to facilitate precisely this kind of two-step processing [33].

The second example is also "cognitive". In the development of "Knowledge Spaces" [6], Doignon and Falmagne call a coherent collection of facts or skills a "knowledge state". These are closed sets which are partially ordered by containment to form a lattice structure [21], which they call a "knowledge space". There is a considerable literature concerning closed knowledge "states" and knowledge "spaces".[5] A somewhat similar approach to cognitive closure was presented in [25].

5.2   Role of Chordless Cycles

Also central to our paper is the concept of "chordless cycles" which constitute the structure of an irreducible trace. Chordless cycles abound in biological organisms as protein polymers.

One example, found in every cell of our bodies, is a 154 node phenylalaninic-glycine-repeat (nuclear pore protein), $\mathcal{N}$, which is shown in Fig. 8.[6] One can easily see the chordless loops, with various linear tendrils attached to them.

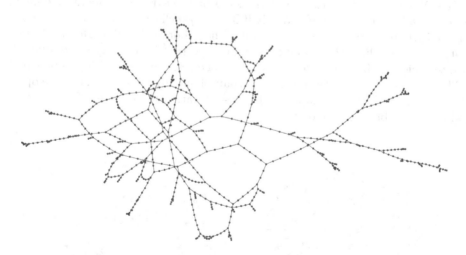

Fig. 8. A 154 node protein polymer

When these are removed by ω, there are 107 remaining elements involved in the chordless cycle structure. These are thought to regulate transport of other proteins across the nuculear membrane [10, 20, 37].

Readily, organisms with any form of memory, *e.g.* "movement toward light yields food", have survival benefit. Nature appears to reuse successful structures. If chordless cycles can successfully regulate one form of transport, it would not be surprising if evolutionary pressure led to their use in other control mechanisms.

Moreover, modification of protein polymers by means of phosphorylation [38] is thought to be involved in short-term memory [14]. For long-term memories, chordless cycles within the dendritic connections between neurons seem more likely.

But is there reason to suspect that memory has any "structural" properties at all?

Perhaps the most important biological memory mechanism is our genetic memory which records the nature of our species. It is known to have a double helix structure which facilitates a near perfect recall. These coded sequences are subsequently "expressed" during development by an expansion process which might be similar to a *non-random* ε.

While the double helix facilitates a reliable read-only memory (ROM); chordless cycles appear facilitate the encoding of eposidic information in a dynamic memory via a process such as ω.

Much of this section is speculation. But, both "closure" [25] and "chordless cycles" [26] would appear to have biological significance. The assertions of this paper have a solid mathematical base [27]. As such, ω and ε provide useful examples within a category of *networks* that can formally model dynamic biological networks. If in addition, they actually model memory consolidation and recall as we suspect, that would be an additional bonus.

A   Appendix

Too much formal mathematics makes a paper hard to read. Yet, it is important to be able to check some of the statements made in the body of the paper. In this appendix we provide a few propositions to formally prove some of our assertions.

The order in which nodes, or more accurately the singleton subsets, of $\mathcal{N}$ are encountered can alter which points are subsumed and subsequently deleted. Nevertheless, we show below that the reduced trace $\mathcal{T} = \mathcal{N}.\omega$ will be unique, up to isomorphism.

Proposition 1. *Let $\mathcal{T} = \mathcal{N}.\omega$ and $\mathcal{T}' = \mathcal{N}.\omega'$ be irreducible subsets of a finite network $\mathcal{N}$, then $\mathcal{T} \cong \mathcal{T}'$.*

Proof. Let $y_0 \in \mathcal{T}$, $y_0 \notin \mathcal{T}'$. Then y_0 can be subsumed by some point y_1 in $\mathcal{T}'$ and $y_1 \notin \mathcal{T}$ else because $y_0.\eta \subseteq y_1.\eta$ implies $y_0 \in \{y_1\}.\varphi$ and $\mathcal{T}$ would not be irreducible.

Similarly, since $y_1 \in \mathcal{T}'$ and $y_1 \notin \mathcal{T}$, there exists $y_2 \in \mathcal{T}$ such that y_1 is subsumed by y_2. So, $y_1.\eta \subseteq y_2.\eta$.

Now we have two possible cases; either $y_2 = y_0$, or not.

Suppose $y_2 = y_0$ (which is often the case), then $y_0.\eta \subseteq y_1.\eta$ and $y_1.\eta \subseteq y_2.\eta$ or $y_0.\eta = y_1.\eta$. Hence $i(y_0) = y_1$ is part of the desired isometry, i.

Now suppose $y_2 \neq y_0$. There exists $y_3 \neq y_1 \in T'$ such that $y_2.\eta \subseteq y_3.\eta$, and so forth. Since T is finite this construction must halt with some y_n. The points $\{y_0, y_1, y_2, \ldots y_n\}$ constitute a complete graph Y_n with $\{y_i\}.\eta = Y_n.\eta$, for $i \in [0, n]$. In any reduction all $y_i \in Y_n$ reduce to a single point. All possibilities lead to mutually isomorphic maps. □

In addition to $N.\omega$ being unique, we may observe that the transformation ω is functional because ω maps all subsets of N onto N_ω. So we can have $\{z\}.\omega = \emptyset$, thus "deleting" z. Similarly, ε is functional because $\emptyset.\varepsilon = \{y\}$ provides for the inclusion of new elements. Both ω and ε are monotone, if we only modify its definition to be $X \subseteq Y$ implies $X.\varepsilon \subseteq Y.\varepsilon$, provided $X \neq \emptyset$.

The following proposition characterizes the structure of irreducible traces.

Proposition 2. *Let N be a finite symmetric network with $T = N.\omega$ being its irreducible trace. If $y \in T$ is not an isolated point then either*

(1) there exists a chordless k-cycle C, $k \geq 4$ such that $y \in C$, or
(2) there exist chordless k-cycles C_1, C_2 each of length ≥ 4 with $x \in C_1$ $z \in C_2$ and y lies on a path from x to z.

Proof.

(1) Let $y \in N_T$. Since y is not isolated, we let $y = y_0$ with $y_1 \in y_0.\eta$, so $(y_0, y_1) \in E$. Since y_1 is not subsumed by y_0, $\exists y_2 \in y_1.\eta$, $y_2 \notin y_0.\eta$, and since y_2 is not subsumed by y_1, $\exists y_3 \in y_2.\eta$, $y_3 \notin y_1.\eta$. Since $y_2 \notin y_0.\eta$, $y_3 \neq y_0$.
  Suppose $y_3 \in y_0.\eta$, then $< y_0, y_1, y_2, y_3, y_0 >$ constitutes a k-cycle $k \geq 4$, and we are done. Suppose $y_3 \notin y_0.\eta$. We repeat the same path extension. $y_3.\eta \not\subseteq y_2.\eta$ implies $\exists y_4 \in y_3.\eta$, $y_4 \notin y_2.\eta$. If $y_4 \in y_0.\eta$ or $y_4 \in y_1.\eta$, we have the desired cycle. If not $\exists y_5, \ldots$ and so forth. Because N is finite, this path extension must terminate with $y_k \in y_i.\eta$, where $0 \leq i \leq n - 3$, $n = |N|$. Let $x = y_0, z = y_k$.
(2) follows naturally. □

Finally, we show that ω preserves the shortest paths between all elements of the trace, T.

Proposition 3. *Let $\sigma(x, z)$ denote a shortest path between x and z in N. Then for all $y \neq x, z, \in \sigma(x, z)$, if y can be subsumed by y', then there exists a shortest path $\sigma'(x, z)$ through y'.*

Proof. We may assume without loss of generality that y is adjacent to z in $\sigma(x, z)$.

Let $< x, \ldots, x_n, y, z >$ constitute $\sigma(x, z)$. If y is subsumed by y', then $y.\eta = \{x_n, y, z\} \subseteq y'.\eta$. So we have $\sigma'(x, z) =< x. \ldots, x_n, y', z >$ of equal length. (Also proven in [23].) □

In other words, z can be removed from $\mathcal{N}$ with the certainty that if there was a path from some node x to z through y, there will still exist a path of equal length from x to z after y's removal.

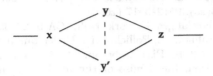

Fig. 9. A network diamond

Figure 9 visually illustrates the situation described in Proposition 3, which we call a **diamond**. There may, or may not, be a connection between y and y' as indicated by the dashed line. If there is, as assumed in Proposition 3, then either y' subsumes y or *vice versa*, depending on the order in which y and y' are encountered by ω. This provides one example of the isomorphism described in Proposition 1. If there is no connection between y and y' then we have two distinct paths between x and z of the same length.

References

1. Braham, C.R., Messaoudi, E.: BDNF function in adult synaptic plasticity: the synaptic consolidation hypothesis. Prog. Neurobiol. **76**(2), 99–125 (2005)
2. Brandes, U.: A faster algorithm for betweeness centrality. J. Math. Sociol. **25**(2), 163–177 (2001)
3. Bretto, A.: Hypergraph Theory: An Introduction. Springer, Cham (2013)
4. Crowder, R.G.: Principles of Learning and Memory. Psychology Press, New York (2015)
5. Davachi, L.: Encoding: the proof is still required. In: Roediger-III, H.L., Dudai, Y., Fitzpatrick, S.M. (eds.) Science of Memory: Concepts, pp. 129–135. Oxford University Press (2007)
6. Doignon, J.P., Falmagne, J.C.: Knowledge Spaces. Springer, Berlin (1999)
7. Farber, M., Jamison, R.E.: Convexity in graphs and hypergraphs. SIAM J. Algebra Discrete Methods **7**(3), 433–444 (1986)
8. Freeman, L.C.: Centrality in social networks, conceptual clarification. Soc. Netw. **1**, 215–239 (1978/1979)
9. Freeman, L.C.: Going the wrong way on a one-way street: centrality in physics and biology. J. Soc. Struct. **9**, 1–15 (2008)
10. Frey, S., Görlich, D.: A saturated FG-repeat hydrogel can reproduce the permeability of nuclear pore complexes. Cell **130**, 512–523 (2007)
11. Gabrieli, J.D.E.: Cognitive neuroscience of human memory. Ann. Rev. Psychol. **49**, 87–115 (1998)
12. Gazzaniga, M.S., Ivry, R.B., Mangun, G.R., Steven, M.S.: Cognitive Neuroscience, The Biology of the Mind. W.W. Norton, New York (2009)
13. Granovetter, M.S.: The strength of weak ties. Am. J. Sociol. **78**(6), 1360–1380 (1973)

14. LeDoux, J.E.: Consolidation: Challenging the traditional view. In: Roediger-III, H.L., Dudai, Y., Fitzpatrick, S.M. (eds.) Science of Memory: Concepts, pp. 171–175. Oxford University Press (2007)

15. Lim, R.Y.H., Huang, N.P., Köser, J., Deng, J., et al.: Flexible phenylalanine-glycine nucleoporins as entropic barriers to nucleocytoplasmic transport. Proc. Nat. Acad. Science (PNAS) **103**(25), 9512–9517 (2006)

16. McKee, T.A.: How chordal graphs work. Bull. ICA **9**, 27–39 (1993)

17. Moshfeghi, Y., Triantafillou, P., Pollick, F.E.: Understanding information need: an fMRI study. In: SIGIR 2016, Pisa, Italy, pp. 335–344, July 2016

18. Nadel, L.: Consolidation: the demise of the fixed trace. In: Roediger-III, H.L., Dudai, Y., Fitzpatrick, S.M. (eds.) Science of Memory: Concepts, pp. 177–181. Oxford University Press (2007)

19. Orlandic, R., Pfaltz, J., Taylor, C.: A functional database representation of large sets of objects. In: Wang, H., Sharaf, M.A. (eds.) ADC 2014. LNCS, vol. 8506, pp. 189–197. Springer, Cham (2014). doi:10.1007/978-3-319-08608-8_17

20. Patel, S.S., Belmont, B.J., Sante, J.M., Rexach, M.F.: Natively unfolded nucleoporins gate protein diffusion across the nuclear pore complex. Cell **129**, 83–96 (2007)

21. Pfaltz, J.L.: Closure lattices. Discrete Math. **154**, 217–236 (1996)

22. Pfaltz, J.L.: Neighborhood expansion grammars. In: Ehrig, H., Engels, G., Kreowski, H.-J., Rozenberg, G. (eds.) TAGT 1998. LNCS, vol. 1764, pp. 30–44. Springer, Heidelberg (2000). doi:10.1007/978-3-540-46464-8_3

23. Pfaltz, J.L.: Finding the mule in the network. In: Alhajj, R., Werner, B. (eds.) International Conference on Advances in Social Network Analysis and Mining, ASONAM 2012, Istanbul, Turkey, pp. 667–672, August 2012

24. Pfaltz, J.L.: The irreducible spine(s) of undirected networks. In: Lin, X., Manolopoulos, Y., Srivastava, D., Huang, G. (eds.) WISE 2013. LNCS, vol. 8181, pp. 104–117. Springer, Heidelberg (2013). doi:10.1007/978-3-642-41154-0_8

25. Pfaltz, J.L.: Using closed sets to model cognitive behavior. In: Ray, T., Sarker, R., Li, X. (eds.) ACALCI 2016. LNCS (LNAI), vol. 9592, pp. 13–26. Springer, Cham (2016). doi:10.1007/978-3-319-28270-1_2

26. Pfaltz, J.L.: A role for chordless cycles in the retrieval and representation of information. In: Proceedings of 6th International Workshop on Querying Graph Structured Data (GraphQ 2017), Venice, IT (2017, to appear)

27. Pfaltz, J.L.: Two network transformations. Math. Appl. **6**(1) (2017, to appear)

28. Phinney, A.L., Calhoun, M.E., Wolfer, D.P., Lipp, H.P., Zheng, H., Jucker, M.: No hippocampal neuron or synaptic bouton loss in learning-impaired aged β-amyloid precursor protein-null mice. Neuroscience **90**(4), 1207–1216 (1999)

29. Purves, D., Augustine, G.J., Fitzpatrick, D., et al.: Neuroscience. Sinauer Assoc., Sunderland (2008)

30. Rolls, E.T., Treves, A.: Neural Networks and Brain Function. Oxford University Press, Oxford (1998)

31. Rosenfeld, A., Pfaltz, J.L.: Sequential operations in digital picture processing. J. ACM **13**(4), 471–494 (1966)

32. Ryan, T.J., Roy, D.S., Pignatelli, M., Arons, A., Tonegawa, S.: Engram cells retain memory under retrograde amnesia. Science **348**(6238), 1007–1013 (2015)

33. Sarti, A., Citti, G., Petitot, J.: Functional geometry of the horizontal connectivity in the primary visual cortex. J. Physiol. Paris **103**(1–2), 37–45 (2009)

34. Sun, Z., Wang, H., Wang, H., Shao, B., Li., J.: Efficient subgraph matching on billion node graphs. In: Proceedings of the VLDB Endowment (originally presented at VLDB Conference, Istanbul, Turkey), vol. 5(9), pp. 788–799 (2012)

35. Tronson, N.C., Taylor, J.R.: Molecular mechanisms of memory reconsolidation. Nat. Rev. Neurosci. **8**(4), 262–275 (2007)
36. Tulving, E.: Coding and representation: searching for a home in the brain. In: Roediger-III, H.L., Dudai, Y., Fitzpatrick, S.M. (eds.) Science of Memory: Concepts, pp. 65–68. Oxford University Press (2007)
37. Weis, K.: The nuclear pore complex: oily spaghetti or gummy bear? Cell **130**, 405–407 (2007)
38. Zhu, F., Guan, Y.: Predicting dynamic signaling network response under unseen perturbations. Bioinformatics **30**(19), 2772–2778 (2014)

Phylogenetics

Inferring the Distribution of Fitness Effects (DFE) of Newly-Arising Mutations Using Samples Taken from Evolving Populations in Real Time

Philip J. Gerrish[1,2,3](✉) and Nick Hengartner[3]

[1] School of Biology, Georgia Institute of Technology,
310 Ferst St, Atlanta, GA 30332, USA
pgerrish@gatech.edu
[2] Instituto de Ciencias Biomédicas, Universidad Autónoma de Ciudad Juárez,
32310 Chihuahua, Mexico
[3] Theoretical Biology and Biophysics, Los Alamos National Laboratory,
MS K710, Los Alamos, NM 87545, USA

Abstract. The DFE characterizes the mutational "input" to evolution, while natural selection largely determines how this input gets sorted into an evolutionary "output". The output cannot contain novel genetic material that is not present in the input and, as such, understanding the DFE and its dynamics is crucial to understanding evolution generally. Despite this centrality to evolution, however, the DFE has remained elusive primarily due to methodological difficulties. Here, we propose and assess a novel framework for estimating the DFE which removes the biasing effects of selection statistically. We propose a statistic for characterizing the difference between two inferred DFEs, taken from two different populations or from the same population at different time points. This allows us to study the evolution of the DFE and monitor for structural changes in the DFE.

Keywords: Adaptive evolution · Fitness mutation · Population genetics · Cumulant expansion · Empirical characteristic function

1 Introduction

Evolution requires genetic variation. Mendelian inheritance provides a mechanism for the *transmission* and *maintenance* of genetic variation, but it does not explain or characterize the *origin* of this variation. Ultimately, the origin of genetic variation is mutation: mutation is what creates the different alleles whose presence in a population can then be maintained through Mendelian inheritance, providing fodder for subsequent adaptive evolution. *Fitness mutations* (mutations that change fitness) have been called the "raw material" or "fuel" of evolution, because they provide the input to natural selection. Despite this essential role in evolution, however, we know almost nothing about the general nature

© Springer International Publishing AG 2017
D. Figueiredo et al. (Eds.): AlCoB 2017, LNBI 10252, pp. 103–114, 2017.
DOI: 10.1007/978-3-319-58163-7_7

of such mutations, or even if they have a general nature. For example, what fraction of such mutations increase fitness? What fraction decrease fitness? Do these have large or small effects on fitness? Generally speaking, what is the distribution of their fitness effects (DFE)? Is there a canonical form for the DFE or do the contingencies outweigh the generalities? How does the DFE change over time, as a population adapts? Given their pivotal role in evolution, it indeed seems surprising that such fundamental questions remain largely unanswered.

In sexual populations, the effects of many mutations can be masked through dominance effects. As a result, a significant fraction of fitness mutations may not be detectable in fitness assays of population samples. Furthermore, deleterious mutations can be eliminated efficiently in sexual populations and their past occurrence may thus be difficult to infer. These features of sexual populations might further confound the already-difficult inverse problem of characterizing the DFE, and such considerations would therefore suggest that attempts to infer the DFE from population samples should start with asexual populations. With this in mind, we here develop methods primarily for use with asexual populations, although the basic framework we develop can be adapted to the case of sexual populations (future work).

1.1   Previous Studies

The primary reasons for the deficiency in our knowledge of the DFE are methodological. Estimation of the DFE requires that two largely contradictory conditions are simultaneously met, namely: (1) that a large sample of candidate replications is surveyed (ideally, a population), and (2) that the strong biasing effects of natural selection are somehow removed. Mutation accumulation assays, for example, remove the biasing effects of selection by reducing the population to the smallest possible effective size; in other words, they achieve (2) above by sacrificing (1). These assays are well-suited to estimating the deleterious mutation rate and the mean deleterious effect of mutations but, because of the restricted numbers of mutants surveyed, not much more information can be extracted.

Largely owing to these methodological difficulties, understanding the DFE remains a key goal of evolutionary genetics [2]. Despite some recent progress based on genome sequencing and experimental evolution approaches [1,4,5,7,8, 12,15,18], much remains to be understood. In particular, the notion that there is a canonical form for the DFE is questionable: it seems intuitively reasonable, for example, that the DFE will be strongly affected by the evolutionary history of a population in its current environment [12].

1.2   Evolution and the DFE

The DFE's Right Tail. A particular feature of the DFE that is of special relevance to evolutionary biology is the right tail. The right tail contains information about the general nature of adaptive mutations. For example, is the probability of acquiring adaptive mutations of large effect vanishingly small, e.g., does the probability of adaptive mutations fall off exponentially as fitness increases? Or is the probability of large-effect adaptive mutations non-negligible, e.g., does the probability fall off as a power law? The answer to these questions will have implications for the general nature of adaptations and evolutionary dynamics, and fundamental questions in evolution such as those surrounding the evolution of sex.

The DFE and the Fitness Landscape. Indirectly, the DFE, its evolution, and particularly its right tail also carry information about the general nature of the underlying adaptive landscape, or at least the immediate neighborhood therein. Adaptive landscapes define the genetic "terrain" that populations navigate through the actions of mutation, recombination, selection and drift. Qualitative differences in the topography of adaptive landscapes can have implications for the relative importance of these four evolutionary factors. A question that has guided some of the work on the DFE to date is: does the DFE reflect an underlying Fisherian adaptive topography [10,11]?

1.3   The Present Study

Fitness mutations observed in the present occurred some time in the past, and inferring the DFE thus requires a dynamical model of fitness evolution. While the history of such models is as old as population genetics itself, much recent progress has been made, especially where asexual and facultatively sexual populations are concerned, and the powerful tools of statistical physics have been applied to evolutionary dynamics with some success [3,6,13,14,16,17,19].

The theoretical framework we develop here provides a way to characterize the DFE based on fitness measurements of individuals drawn at random from a population; it does this by statistically accounting for the biasing effects of natural selection. In a forthcoming article, we apply the methods developed here to real data from evolving *E. coli* populations. This study lays the theoretical groundwork that, when combined with data from laboratory evolution experiments (a forthcoming paper), will provide what we believe to be the first truly *in situ* look at the DFE of *spontaneously arising* mutations, and how it changes as populations adapt to novel environments.

Our methods provide a way to compute the moments of the DFE, and it is known [9] that moments carry plenty of information about distribution tails. Indeed, the simulation studies we describe here confirm this: using simulated data, our methods give a very accurate reconstruction of the right tail of the DFE. As mentioned in the previous subsection, this tail is of paramount relevance to the general nature of evolution.

Characterizing the *Evolution* of the DFE. Structural changes in the DFE and/or the appearance of non-stationarity in its evolution might provide early warning signs of impending evolutionary "shifts". Two examples of such shifts might be: (1) metastatic transitions in cancer, and (2) zoonotic events such as avian flu's evolutionary leap to infect a new animal host. We illustrate how our methods can be applied to detect structural changes in the DFE.

2 Fitness Evolution

2.1 Evolution of the Distribution of Fitnesses

Let $u(x,t)$ denote probability density in fitness x at time t (i.e., $\int_x u(x,t) = 1$ for all t) for an evolving population. Under selection and mutation, u evolves as:

$$\partial_t u(x,t) = (x - \bar{x})u(x,t) + U\int_\gamma u(x-\gamma,t)g(\gamma,t) - Uu(x,t)$$

where U is genomic mutation rate, and $g(\gamma,t)$ is probability density for fitness changes incurred by mutation (i.e., the DFE).

2.2 Evolution of the Corresponding Characteristic Function

Let $C(\varphi,t)$ denote the characteristic function for $u(x,t)$, i.e., $C(\varphi,t) = \int e^{i\varphi x}u(x,t)dx$ and let $M(\varphi,t)$ denote the characteristic function for the DFE, i.e., $M(\varphi,t) = \int e^{i\varphi x}g(x,t)dx$. The transformed equation is:

$$\partial_t C(\varphi,t) = -i\partial_\varphi C(\varphi,t) + i\partial_\varphi C(0,t)C(\varphi,t) + UC(\varphi,t)(M(\varphi,t) - 1).$$

Over the time interval in question (assumed to be short on evolutionary time scales), we will suppose the DFE is invariant such that $M(\varphi,t) = M(\varphi)$. Let $\Phi(\varphi,t) = \ln C(\varphi,t)$; then we have:

$$\partial_t \Phi(\varphi,t) = -i\partial_\varphi \Phi(\varphi,t) + i\partial_\varphi \Phi(0,t) + U(M(\varphi) - 1).$$

This equation is a variant of the transport equation and, when boundary condition $\Phi(0,t) = 0$ is applied, it has solution:

$$\Phi(\varphi,t) = \Phi(\varphi - it,0) - \Phi(-it,0) - iU\int_{\varphi-it}^{\varphi}(M(\gamma)-1)d\gamma + iU\int_{-it}^{0}(M(\gamma)-1)d\gamma$$

which reduces to:

$$\Phi(\varphi,t) = \Phi(\varphi - it,0) - \Phi(-it,0) - iU\int_{-it}^{0}(M(\varphi+\gamma) - M(\gamma))d\gamma \qquad (1)$$

2.3 Dynamics of Mean Fitness

Mean fitness at time t, denoted $\bar{x}(t)$, is derived as follows:

$$\bar{x}(t) = (-i)\frac{\partial}{\partial\varphi}[\Phi(\varphi,t)]_{\varphi=0}$$

Letting $\Phi_0(\varphi) = \Phi(\varphi,0)$, this gives a general expression for mean fitness evolution under selection and mutation:

$$\bar{x}(t) = -i\Phi_0'(-it) + U[M(-it) - 1] \qquad (2)$$

where the prime denotes derivative.

2.4   Dynamics of Fitness Variance and Higher Cumulants

Fitness variance at time t, denoted $\sigma_x^2(t)$, is derived as follows:

$$\sigma_x^2(t) = (-i)^2 \frac{\partial^2}{\partial \varphi^2} \left[\Phi(\varphi, t)\right]_{\varphi=0}$$

giving:

$$\sigma_x^2(t) = -\Phi_0''(-it) - iU[M'(-it) - M'(0)] \tag{3}$$

And generally, the jth cumulant at time t, denoted $\kappa_j(t)$, is given by:

$$\kappa_j(t) = (-i)^j \frac{\partial^j}{\partial \varphi^j} \left[\Phi(\varphi, t)\right]_{\varphi=0} \tag{4}$$

2.5   Connection to Classical Theory

Without mutation, the equation for mean fitness evolution is:

$$\bar{x}(t) = -i\Phi_0'(-it)$$

which may be rewritten as:

$$\bar{x}(t) = -i \frac{\partial}{\partial \varphi} \left[i\kappa_1\varphi - \frac{1}{2}\kappa_2\varphi^2 - i\frac{1}{6}\kappa_3\varphi^3 + \dots \right]_{\varphi=-it}$$
$$= -i \left[i\kappa_1 + \kappa_2 it + \mathcal{O}(t^2) + \dots \right]$$

Noting that $\kappa_1 = \bar{x}(0)$ and $\kappa_2 = \sigma_x^2(0)$, and computing for some very small time increment into the future, $t = dt$, we have:

$$\bar{x}(dt) = \bar{x}(0) + \sigma_x^2(0)dt + \mathcal{O}(dt^2), \quad \text{or}$$
$$\frac{\bar{x}(dt) - \bar{x}(0)}{dt} = \sigma_x^2(0) + \mathcal{O}(dt)$$

And letting $dt \to 0$ gives:

$$\frac{d\bar{x}}{dt} = \sigma_x^2$$

which is the continuous-time formulation of Fisher's fundamental theorem of natural selection.

3   Parametric Estimation of the DFE

From the above developments, we derive two methods for parametric estimation of the DFE, each with different merits and shortcomings. We will call these methods the "integral" and "derivative" methods.

3.1   Integral Method

We define the functions:

$$G(\varphi, t) = iU \int_{-it}^{0} (M(\varphi + \gamma) - M(\gamma))d\gamma,$$

and

$$F(\varphi, t) = \Phi(\varphi - it, 0) - \Phi(-it, 0) - \Phi(\varphi, t).$$

Then Eq. (1) may be rewritten as:

$$G(\varphi, t) = F(\varphi, t).$$

Now we define the parametric counterpart to $G(\varphi, t)$:

$$\tilde{G}(\varphi, t) = iU \int_{-it}^{0} (P(\varphi + \gamma) - P(\gamma))d\gamma,$$

where $P(\varphi)$ is the CF of a known distribution chosen to represent the DFE with parameter vector $\boldsymbol{\theta}$. It will often be the case that $\tilde{G}(\varphi, t)$ has a closed-form expression when $P(\varphi)$ is analytic; these cases are convenient but strictly speaking we are not limited to such cases. Next, we define the empirical counterpart to $F(\varphi, t)$:

$$\tilde{F}(\varphi, t) = \tilde{\Phi}(\varphi - it, 0) - \tilde{\Phi}(-it, 0) - \tilde{\Phi}(\varphi, t).$$

where

$$\tilde{\Phi}(\varphi, t) = \log \sum_{j} e^{i\varphi X_j(t)},$$

the empirical log CF (or ELCF), computed by simply inserting empirically-determined fitness measurements $X_j(t)$ taken at time t.

We let τ denote the amount of time between two sampling time points. The parameters of the DFE and the genomic mutation rate are then estimated by finding the parameters $\hat{\boldsymbol{\theta}}$ and $\hat{U}$ that minimize the quantity:

$$\left(\tilde{G}(\varphi, \tau) - \tilde{F}(\varphi, \tau) \right)^2.$$

Operationally, $\hat{\boldsymbol{\theta}}$ and $\hat{U}$ are determined by minimizing $\sum_j (\Re \tilde{G}(\varphi_j, \tau) - \Re \tilde{F}(\varphi_j, \tau))^2 + \sum_j (\Im \tilde{G}(\varphi_j, \tau) - \Im \tilde{F}(\varphi_j, \tau))^2$ for a finite (small) set of values φ_j, or by minimizing $\int_{\varphi} (\Re \tilde{G}(\varphi, \tau) - \Re \tilde{F}(\varphi, \tau))^2 w(\varphi) + \int_{\varphi} (\Im \tilde{G}(\varphi, \tau) - \Im \tilde{F}(\varphi, \tau))^2 w(\varphi)$, where $w(\varphi)$ is a weighting function that typically gives more weight to values of φ near zero.

3.2   Derivative Method of DFE Estimation

The foregoing "integral" method of DFE estimation is optimal in the sense that no information is lost. In practice, however, it can be a bit unwieldy and computationally slow if $\tilde{G}(\varphi, \tau)$ cannot be expressed in closed form. Furthermore, we have found that the above method can sometimes be numerically unstable. For this reason, at the risk of stating the obvious, we now present the method obtained immediately from the foregoing method by simply taking its derivative.

Assuming that $P(\varphi)$ is analytic, we have:

$$\partial_\varphi \tilde{G}(\varphi, \tau) = iU(P(\varphi) - P(\varphi - i\tau)),$$

and:

$$\partial_\varphi \tilde{F}(\varphi, \tau) = \partial_\varphi \tilde{\Phi}(\varphi - i\tau, 0) - \partial_\varphi \tilde{\Phi}(\varphi, \tau).$$

Here, parameters of the DFE and the genomic mutation rate are estimated by finding the parameters $\hat{\boldsymbol{\theta}}$ and $\hat{U}$ that minimize the quantity:

$$\left(\partial_\varphi \tilde{G}(\varphi, \tau) - \partial_\varphi \tilde{F}(\varphi, \tau)\right)^2.$$

Again, $\hat{\boldsymbol{\theta}}$ and $\hat{U}$ are operationally determined by minimizing $\sum_j (\Re \partial_\varphi \tilde{G}(\varphi_j, \tau) - \Re \partial_\varphi \tilde{F}(\varphi_j, \tau))^2 + \sum_j (\Im \partial_\varphi \tilde{G}(\varphi_j, \tau) - \Im \partial_\varphi \tilde{F}(\varphi_j, \tau))^2$ for a finite (small) set of values φ_j, or by minimizing $\int_\varphi (\Re \partial_\varphi \tilde{G}(\varphi, \tau) - \Re \partial_\varphi \tilde{F}(\varphi, \tau))^2 w(\varphi) + \int_\varphi (\Im \partial_\varphi \tilde{G}(\varphi, \tau) - \Im \partial_\varphi \tilde{F}(\varphi, \tau))^2 w(\varphi)$.

Multiplying both $\partial_\varphi \tilde{G}(\varphi, \tau)$ and $\partial_\varphi \tilde{F}(\varphi, \tau)$ by i, the quantity to be minimized may be written:

$$\left(U[P(\varphi) - P(\varphi - i\tau)] - i[\partial_\varphi \tilde{\Phi}(\varphi, \tau) - \partial_\varphi \tilde{\Phi}(\varphi - i\tau, 0)]\right)^2$$

which has a curious symmetry.

4   Non-parametric Distance Measures for DFEs

The aim here is to develop a statistic that could be used as a measure of discrepancy between two inferred DFEs, and as a tool for statistical inference of such discrepancy.

4.1   Conjecture

The DFE is well-characterized by the function:

$$G(\varphi, t) = iU \int_{-it}^{0} (M(\varphi + \gamma) - M(\gamma)) d\gamma.$$

Thus, changes in the DFE may be detected as changes in $\tilde{G}(\varphi)$, which may be estimated from population samples of fitness as:

$$\tilde{G}(\varphi) = \left[\tilde{\Phi}(\varphi - i\tau, 0) - \tilde{\Phi}(-i\tau, 0) - \tilde{\Phi}(\varphi, \tau)\right] w(\varphi),$$

where $w(\varphi)$ is a weighting function. In what follows, we will generally use gaussian weighting, $w(\varphi) = \mathcal{N}_{h\varphi}(0,1) = e^{-(h\varphi)^2/2}/\sqrt{2\pi}$, where $1/h$ determines the bandwidth.

Time $t = 0$ may be understood as the time at which the first sample is taken (beginning of a sampling interval) and $t = \tau$ may be understood as the time at which the second sample is taken (end of the same sampling interval).

In light of the foregoing conjecture, our objective is now to define a measure of distance between $G_1(\varphi)$ as inferred from one pair of sampling time points τ generations apart and a different $G_2(\varphi)$ as inferred from a different pair of sampling time points τ generations apart. (Note: Here we focus on DFEs inferred from pairs of data points. In practice, however, G_1 and G_2 may both be averages over several pairs of data points.) This distance measure is taken to be an indication of distance between the corresponding DFEs.

4.2   Mahalanobis-Based Distance Measures

Bootstrap Method. We define a set of points along the φ-axis, denoted φ_k, $k = 1, 2, ..., n$. The first step is to generate a large set of ELCFs, $\tilde{\Phi}^b(\varphi, 0)$ and $\tilde{\Phi}^b(\varphi, \tau)$, by resampling the $X_j(0)$ and $X_j(\tau)$, respectively, that were used to compute $G_1(\varphi, \tau)$. For each pair $\tilde{\Phi}^b(\varphi, 0)$ and $\tilde{\Phi}^b(\varphi, \tau)$, we compute

$$\tilde{G}^b(\varphi) = \left[\tilde{\Phi}^b(\varphi - i\tau, 0) - \tilde{\Phi}^b(-i\tau, 0) - \tilde{\Phi}^b(\varphi, \tau)\right] w(\varphi)$$

from which we compute the vector:

$$g_b = (\Re\tilde{G}^b(\varphi_1), ..., \Re\tilde{G}^b(\varphi_n), \Im\tilde{G}^b(\varphi_1), ..., \Im\tilde{G}^b(\varphi_n)). \tag{5}$$

The covariances among the g_b define the covariance matrix Ω_1.

First, we define corresponding vectors g_1 and g_2 as defined by (5), but using all the data (not bootstrapped) at time points 1 and 2. The bootstrapped distance statistic is:

$$D_b = (g_2 - g_1)^T \Omega_1^{-1} (g_2 - g_1) \tag{6}$$

where the T superscript denotes transposition, and the -1 superscript denotes inverse.

Analytical Method. Define:

$$C_{rr} = \frac{1}{2}[\Re G_1(\varphi_j - \varphi_k) + \Re G_1(\varphi_j + \varphi_k)] - \Re G_1(\varphi_j)\Re G_1(\varphi_k) \quad \forall j, k$$

$$C_{ri} = \frac{1}{2}[\Im G_1(\varphi_j - \varphi_k) + \Im G_1(\varphi_j + \varphi_k)] - \Re G_1(\varphi_j)\Im G_1(\varphi_k) \quad \forall j, k$$

$$C_{ii} = \frac{1}{2}[\Re G_1(\varphi_j - \varphi_k) - \Re G_1(\varphi_j + \varphi_k)] - \Im G_1(\varphi_j)\Im G_1(\varphi_k) \quad \forall j, k$$

Then, the covariance matrix to be used is:

$$C_1 = \begin{pmatrix} C_{rr} & C_{ri} \\ C_{ri} & C_{ii} \end{pmatrix}$$

The analytically-defined distance measure is thus defined as:

$$D_a = (g_2 - g_1)^T C_1^{-1} (g_2 - g_1)$$

4.3   Norm-Based Distance Measures

The practicality of the foregoing Mahalanobis-based distance is limited by our observation that the covariance matrices can often have undefined inverses. As such, we have experimented with the following alternative distance measures:

$$d_A(\tilde{G}_1, \tilde{G}_2) = \int_D |\tilde{G}_2(\varphi) - \tilde{G}_1(\varphi)| d\varphi \tag{7}$$

$$d_B(\tilde{G}_1, \tilde{G}_2) = \int_D [\tilde{G}_2(\varphi) - \tilde{G}_1(\varphi)]^2 d\varphi \tag{8}$$

$$d_C(\tilde{G}_1, \tilde{G}_2) = \int_D \left[\sqrt{\tilde{G}_2(\varphi)} - \sqrt{\tilde{G}_1(\varphi)} \right]^2 d\varphi \tag{9}$$

where D is some finite interval whose width is proportional to $1/h$. In general, we used $\tilde{G}_1$ to generate the null by bootstrapping. We thus determined p-values for the above distance measures as follows:

$$p_{1,2} = \mathbb{P}[d(\tilde{G}_1, \tilde{G}_2) > d(\tilde{G}_1, \tilde{G}_1^b)] \tag{10}$$

5   Simulation Tests

5.1   Fitness Dynamics

Our first step was to test the accuracy of theoretical predictions for fitness dynamics, given by Eq. (2). To this end, we performed fully stochastic, individual-based, Wright-Fisher-with-Mutation (WFM) simulations and compared the trajectory of mean fitness from those simulations with that predicted by Eq. (2). In these simulations, fitnesses of newly-arising mutations were drawn at random from a known DFE. The comparison is plotted in Fig. 1.

5.2   Reconstructing the DFE and Estimating U

We sought to assess the accuracy with which our methods can: (1) reconstruct the DFE, given sample sizes on par with our experimental sample sizes, and (2) estimate genomic mutation rate U. To this end, we performed Wright-Fisher-with-Mutation (WFM) simulations. In these simulations, fitnesses of newly-arising mutations were drawn at random from a known DFE. We reconstructed the DFE and estimated U using the methods described above, based samples of 100 fitnesses drawn at random from the evolving population every $\tau = 14$ generations (to mimic our experimental setup). We have found that our parametric methods can fairly accurately estimate U and reconstruct the DFE, even when we purposely choose the wrong parametric form.

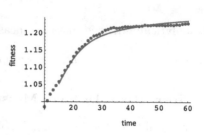

Fig. 1. Comparing theoretical predictions of Eq. (2) with fully stochastic, individual-based, WFM simulations. Blue dots represent means of fitnesses and phenotypes in a simulation; red curves represent theoretical predictions based on a single sample of size 100 drawn at random from the simulated population at generation 15 and inserted into Eq. (2). Population size for the simulations was 5000. Mutation rate was $U = 0.01$; the DFE was a two-sided exponential distribution with mean 0.03 on both sides and weights 0.99 and 0.01 assigned to the deleterious and beneficial sides, respectively; fitnesses in the initial population (at $t = 0$) were drawn at random from a normal distribution with mean one and standard deviation 0.05. (Color figure online)

5.3   Reconstructing DFE Dynamics

We further sought to assess the sensitivity with which our methods can detect changes in the DFE over time. Again, we performed WFM simulations, but in these simulations the known DFE changes over time, either slowly or abruptly. One example of these comparisons is illustrated in Fig. 3 which plots a heat map of inferred DFE pairwise distances. A fitness "ripple" was created in these simulations by simply changing the shape of the DFE between generations 360 and 380. While based on only a subsample (sample size 200) taken at the population level, the source of this ripple was accurately detected by our methods. Figure 2 shows the p-values calculated for a simulation in which the DFE remained constant until generation 2000 at which point it underwent an abrupt qualitative change. The p-values reflect the probability that the DFE is the same as it was at generation 1000; this probability becomes small after generation 2000, when the known DFE changed.

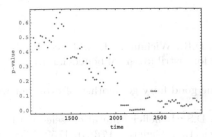

Fig. 2. *p*-values for non-parametric rejection of the null hypothesis that the inferred DFE is not different from the DFE inferred at generation 1000, employing distance measure (7). Distances were computed from sample sizes of 200 fitness drawn at random from simulations of evolving populations of size 10, 000. In simulations, the DFE undergoes an abrupt, qualitative change at generation 2000. This change is detected from our methods by the falling *p*-values after generation 2000.

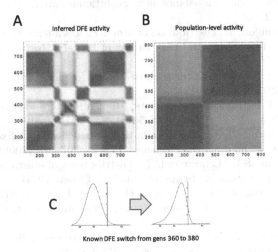

Fig. 3. A simulation test of our methods to detect changes in the DFE. (**A**) Heat map of the distance statistic for pairs of inferred DFE's, given by Eq. (6), based on samples of 200 fitness drawn at random from the evolving population at $\tau = 20$ generation intervals. (**B**) Heat map of a similar statistic for pairs of observed fitness distributions. (**C**) In the simulations, a "fitness ripple" was created by imposing a change in the shape of the DFE from a normal to a Gumbel distribution between generations 360 and 380; while the shape of the distribution changed, the mean and the probability mass above zero was kept constant.

Acknowledgements. We thank Guillaume Martin, Thomas Burr, Paul Sniegowski, Tanya Singh, Thomas Bataillon, and Eduarda Pimentel for helpful discussions, and three anonymous referees. PG carried out some of this work while visiting Aarhus University, Denmark. PG received financial support from NASA grant NNA15BB04A.

References

1. Beisel, C.J., Rokyta, D.R., Wichman, H.A., Joyce, P.: Testing the extreme value domain of attraction for distributions of beneficial fitness effects. Genetics **176**(4), 2441–2449 (2007)
2. Charlesworth, B.: The good fairy godmother of evolutionary genetics. Curr. Biol. CB **6**(3), 220 (1996)
3. Desai, M.M., Fisher, D.S.: Beneficial mutation selection balance and the effect of linkage on positive selection. Genetics **176**(3), 1759–1798 (2007)
4. Elena, M.: Rate of deleterious mutation and the distribution of its effects on fitness in vesicular stomatitis virus. J. Evol. Biol. **12**(6), 1078–1088 (1999)
5. Elena, S.F., Ekunwe, L., Hajela, N., Oden, S.A., Lenski, R.E.: Distribution of fitness effects caused by random insertion mutations in Escherichia coli. Genetica **102–103**(1–6), 349–358 (1998)
6. Goyal, S., Balick, D.J., Jerison, E.R., Neher, R.A., Shraiman, B.I., Desai, M.M.: Dynamic mutation-selection balance as an evolutionary attractor. Genetics **191**(4), 1309–1319 (2012)
7. Kassen, R., Bataillon, T.: Distribution of fitness effects among beneficial mutations before selection in experimental populations of bacteria. Nat. Genet. **38**(4), 484–488 (2006)
8. Keightley, P.D.: Rates and fitness consequences of new mutations in humans. Genetics **190**(2), 295–304 (2012)
9. Lindsay, B.G., Prasanta Basak Associate: Moments determine the tail of a distribution (but not much else). Am. Stat. **54**(4), 248–251 (2000)
10. Martin, G., Lenormand, T.: The distribution of beneficial and fixed mutation fitness effects close to an optimum. Genetics **179**(2), 907–916 (2008)
11. Martin, G., Elena, S.F., Lenormand, T.: Distributions of epistasis in microbes fit predictions from a fitness landscape model. Nat. Genet. **39**(4), 555–560 (2007)
12. McDonald, M., Cooper, T.F., Beaumont, H., Rainey, P.: The distribution of fitness effects of new beneficial mutations in Pseudomonas fluorescens. Biol. Lett. **7**, 98–100 (2011)
13. Neher, R.A., Shraiman, B.I.: Genetic draft and quasi-neutrality in large facultatively sexual populations. Genetics **188**(4), 975–996 (2011)
14. Neher, R.A.: Genetic draft, selective interference, and population genetics of rapid adaptation. Ann. Rev. Ecol. Evol. Syst. **44**(1), 195–215 (2013)
15. Perfeito, L., Fernandes, L., Mota, C., Gordo, I.: Adaptive mutations in bacteria: high rate and small effects. Science **317**, 813–815 (2007). (New York, NY)
16. Rouzine, I.M., Brunet, E., Wilke, C.O.: The traveling-wave approach to asexual evolution: Muller's ratchet and speed of adaptation. Theor. Popul. Biol. **73**(1), 24–46 (2008)
17. Rouzine, I.M., Wakeley, J., Coffin, J.M.: The solitary wave of asexual evolution. PNAS **100**(2), 587–592 (2003)
18. Sanjuán, R., Moya, A., Elena, S.F.: The distribution of fitness effects caused by single-nucleotide substitutions in an RNA virus. Proc. Nat. Acad. Sci. **101**, 8396–8401 (2004)
19. de Vladar, H.P., Barton, N.H.: The contribution of statistical physics to evolutionary biology. Trends Ecol. Evol. **26**, 424–432 (2011)

An Efficient Algorithm for the Rooted Triplet Distance Between Galled Trees

Jesper Jansson[1,2], Ramesh Rajaby[3,4], and Wing-Kin Sung[3,5(✉)]

[1] Laboratory of Mathematical Bioinformatics, ICR,
Kyoto University, Gokasho, Uji, kyoto 611-0011, Japan
jj@kuicr.kyoto-u.ac.jp
[2] Department of Computing, The Hong Kong Polytechnic University,
Hung Hom, Kowloon, Hong Kong, China
[3] School of Computing, National University of Singapore,
13 Computing Drive, Singapore 117417, Singapore
e0011356@u.nus.edu, ksung@comp.nus.edu.sg
[4] NUS Graduate School for Integrative Sciences and Engineering,
National University of Singapore, 28 Medical Drive, Singapore 117456, Singapore
[5] Genome Institute of Singapore, 60 Biopolis Street, Genome,
Singapore 138672, Singapore

Abstract. The previously fastest algorithm for computing the rooted triplet distance between two input galled trees (i.e., phylogenetic networks whose cycles are vertex-disjoint) runs in $O(n^{2.687})$ time, where n is the cardinality of the leaf label set. Here, we present an $O(n \log n)$-time solution. Our strategy is to transform the input so that the answer can be obtained by applying an existing $O(n \log n)$-time algorithm for the simpler case of two phylogenetic trees a constant number of times.

Keywords: Phylogenetic network comparison · Galled tree · Rooted triplet · Algorithm · Computational complexity

1 Introduction

Measuring the similarity between phylogenetic trees is essential for evaluating the accuracy of methods for phylogenetic reconstruction [11]. The *rooted triplet distance* [5] between two rooted phylogenetic trees having the same leaf label sets is given by the number of phylogenetic trees of size three that are embedded subtrees in either one of the input trees, but not the other. Since two phylogenetic trees with a lot of branching structure in common will typically share many such subtrees, the rooted triplet distance provides a natural measure of how dissimilar the two trees are.

A naive algorithm can compute the rooted triplet distance between two input rooted phylogenetic trees in $O(n^3)$ time, where n is the cardinality of the leaf label set, by directly checking each of the $\binom{n}{3}$ different cardinality-3 subsets of the leaf label set. More involved algorithms have been developed [1,2,4,10], and the asymptotically fastest one [2] solves the problem in $O(n \log n)$ time.

© Springer International Publishing AG 2017
D. Figueiredo et al. (Eds.): AlCoB 2017, LNBI 10252, pp. 115–126, 2017.
DOI: 10.1007/978-3-319-58163-7_8

Gambette and Huber [6] extended the rooted triplet distance from the phylogenetic tree setting to the *phylogenetic network* setting. In a *phylogenetic network* [8,12], internal nodes are allowed to have more than one parent. Phylogenetic networks enable scientists to represent more complex evolutionary relationships than phylogenetic trees, e.g., involving horizontal gene transfer events, or to visualize conflicting branching structure among a collection of two or more phylogenetic trees. The special case of a phylogenetic network in which all underlying cycles are vertex-disjoint is called a *galled tree* [7,8,13]. Galled trees may be sufficient in cases where a phylogenetic tree is not good enough but it is known that only a few reticulation events have happened; see Fig. 9.22 in [8] for a biological example. For a summary of other distances for comparing two galled trees (the Robinson-Foulds distance, the tripartitions distance, the μ-distance, the split nodal distance, etc.), see [3].

The fastest known algorithm for computing the rooted triplet distance between two galled trees relies on triangle counting and runs in $O(n^{2.687})$ time [9]. More precisely, its time complexity is $O(n^{(3+\omega)/2})$, where ω is the exponent in the running time of the fastest existing method for matrix multiplication. Since $\omega < 2$ is impossible, the running time for computing the rooted triplet distance between two galled trees using the algorithm from [9] will never be better than $O(n^{2.5})$. In this paper, we present an algorithm for the case of galled trees that does not use triangle counting but instead transforms the input to an appropriately defined set of *phylogenetic trees* to which the $O(n \log n)$-time algorithm of [2] is applied a constant number of times. Basically, in any galled tree, removing one of the two edges leading to an indegree-2 vertex in every cycle yields a tree which still contains most of the branching information, and we show how to compensate for what is lost by doing so while avoiding double-counting. The resulting time complexity of our new algorithm is $O(n \log n)$.

The paper is organized as follows. Section 2 defines the problem formally, Sect. 3 presents the algorithm and its analysis, and Sect. 4 contains some concluding remarks.

2   Problem Definitions

We recall the following definitions from [9].

A (rooted) *phylogenetic tree* is an unordered, rooted tree in which every internal node has at least two children and all leaves are distinctly labeled. A (rooted) *phylogenetic network* is a directed acyclic graph with a single root vertex and a set of distinctly labeled leaves, and no vertices having both indegree 1 and outdegree 1. A *reticulation vertex* in a phylogenetic network is any vertex of indegree greater than 1. For any phylogenetic network N, define *its underlying undirected graph* as the undirected graph obtained by replacing every directed edge in N by an undirected edge. A *cycle* C in a phylogenetic network is any subgraph with at least three edges whose corresponding subgraph in the underlying undirected graph is isomorphic to a cycle, and the vertex of C that is an ancestor of all vertices on C is called the *root* of C. A phylogenetic network is called a *galled tree* if all of its cycles are vertex-disjoint [7,8,13]. Note that every reticulation vertex in a galled tree must have indegree 2. Every cycle C in a galled tree (also called a *gall*) has exactly one root (also referred to as its *split vertex*) and one reticulation vertex, and C consists of two directed, internally disjoint paths from its split vertex to its reticulation vertex.

A phylogenetic tree with exactly three leaves is called a *rooted triplet*. A rooted triplet leaf-labeled by $\{a,b,c\}$ with one internal node is called a *fan triplet* and is denoted by $a|b|c$, while a rooted triplet leaf-labeled by $\{a,b,c\}$ with two internal nodes is called a *resolved triplet*; in the latter case, there are three possible topologies, denoted by $ab|c$, $ac|b$, and $bc|a$, corresponding to when the lowest common ancestor of the two leaves labeled by a and b, or a and c, or b and c, respectively, is a proper descendant of the root. Let a,b,c be three leaf labels in a phylogenetic network N. The fan triplet $a|b|c$ is *consistent with N* if and only if N contains a vertex v and three directed paths from v to a, from v to b, and from v to c that are vertex-disjoint except for in the common start vertex v. Similarly, the resolved triplet $ab|c$ is *consistent with N* if and only if N contains two vertices v and w ($v \neq w$) such that there are four directed paths of non-zero length from v to a, from v to b, from w to v, and from w to c that are vertex-disjoint except for in the vertices v and w. For any phylogenetic network N, $t(N)$ denotes the set of all rooted triplets (i.e., fan triplets as well as resolved triplets) that are consistent with N.

Definition 1 *(Adapted from [6]). Let N_1, N_2 be two phylogenetic networks on the same leaf label set L. The* rooted triplet distance between N_1 and N_2, *denoted by $d_{rt}(N_1, N_2)$, is the number of fan triplets and resolved triplets with leaf labels from L that are consistent with exactly one of N_1 and N_2.*

(See also Sect. 3.2 in [9] for a discussion of the above definition.) Define $fcount(N_1, N_2)$ as the number of fan triplets consistent with both N_1 and N_2, $rcount(N_1, N_2)$ as the number of resolved triplets consistent with both N_1 and N_2, and $count(N_1, N_2) = fcount(N_1, N_2) + rcount(N_1, N_2)$. Note that for $i \in \{1,2\}$, we have $|t(N_i)| = count(N_i, N_i)$. Then one can compute $d_{rt}(N_1, N_2)$ by the formula $d_{rt}(N_1, N_2) = count(N_1, N_1) + count(N_2, N_2) - 2 \cdot count(N_1, N_2)$. The following result was shown by Brodal *et al.* in [2]:

Theorem 2. [2] *If T_1, T_2 are two phylogenetic trees on the same leaf label set L then $fcount(T_1, T_2)$ and $rcount(T_1, T_2)$ (and hence, $d_{rt}(T_1, T_2)$) can be computed in $O(n \log n)$ time, where $n = |L|$.*

From now on, we assume that the input consists of two galled trees N_1 and N_2 over a leaf label set L and that the objective is to compute $d_{rt}(N_1, N_2)$. We define $n = |L|$. It is known that $d_{rt}(N_1, N_2)$ can be computed in $O(n^{2.687})$ time [9]. Below, we show how to do it faster by using Theorem 2, which yields our main result:

Theorem 3. *If N_1, N_2 are two galled trees on the same leaf label set L then $fcount(N_1, N_2)$ and $rcount(N_1, N_2)$ (and hence, $d_{rt}(N_1, N_2)$) can be computed in $O(n \log n)$ time, where $n = |L|$.*

3   The New Algorithm

Section 3.1 describes how to compute $rcount(N_1, N_2)$ efficiently, while Sect. 3.2 is focused on $fcount(N_1, N_2)$. (Both subsections rely on Theorem 2.) In addition to the definitions provided in Sect. 2, the following notation and terminology will be needed.

Suppose that N is a galled tree. For each internal vertex in N, fix some arbitrarily left-to-right ordering of its children. Then $N^\searrow$ is the tree obtained by removing the right parent edge of every reticulation vertex in N and contracting every edge (if any) leading to a vertex with exactly one child. Similarly, $N^\swarrow$ is the tree formed by removing the left parent edge of every reticulation vertex in N and contracting all edges leading to degree-1 vertices. Let $N^\downarrow$ be the tree formed by removing both the left and right parent edges of the reticulation vertex h in each gall, inserting a new edge between the gall's split vertex and h, and contracting all edges leading to degree-1 vertices.

Let $r(N)$ denote the root of N and let $gall(N)$ be the set of all galls in N. For each $Q \in gall(N)$, let $r(Q)$ be the root of Q and h_Q the reticulation vertex of Q. Let Q_L and Q_R be the *left and right paths of* Q, obtained by removing $r(Q)$, h_Q, and all edges incident to $r(Q)$ and h_Q. A rooted triplet in $t(N)$ with leaf label set $\{x, y, z\}$ is called *ambiguous* if N contains a gall Q such that:

1. x, y, z are in three different subtrees attached to Q or $r(Q)$;
2. exactly one leaf is in the subtree attached to h_Q; and
3. at least one leaf is in a subtree attached to Q_L or Q_R.

The ambiguous triplets are partitioned into type-A, type-B, and type-C triplets, defined as follows (see Fig. 1 for an illustration):

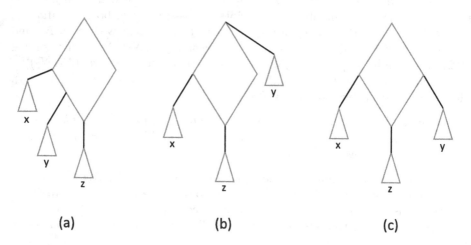

(a)   (b)   (c)

Fig. 1. (a), (b) and (c) illustrate the definitions of type-A, type-B, and type-C triplets, respectively.

- $\{x, y, z\}$ is a *type-A triplet of N* if there exists a gall Q in N such that two leaves in $\{x, y, z\}$ appear in two different subtrees attached to the same Q_δ ($\delta = L$ or R) while the remaining leaf appears in the subtree rooted at h_Q. Furthermore, if $\{x, y, z\}$ is a type-A triplet of N and $\{x, y, z\}$ are attached to a gall Q in N with z appearing in the subtree rooted at h_Q then $\{x, y, z\}$ is called a type-A triplet of N with *reticulation leaf z*.
- $\{x, y, z\}$ is a *type-B triplet of N* if there exists a gall Q in N such that, among the three leaves in $\{x, y, z\}$, one leaf is attached to $r(Q)$ but is not in Q, another leaf appears in a subtree attached to Q_L (or Q_R) and the last leaf appears in the subtree rooted at h_Q.
- $\{x, y, z\}$ is a *type-C triplet of N* if there exists a gall Q in N such that, among the three leaves in $\{x, y, z\}$, one leaf appears in a subtree attached to Q_L, another leaf appears in a subtree attached to Q_R, and the last leaf appears in the subtree rooted at h_Q.

3.1   Counting Common Resolved Triplets in N_1 and N_2

The main idea of our algorithm for computing $rcount(N_1, N_2)$ is to count the common resolved triplets between N_1 and each of the three trees $N_2^{\swarrow}$, $N_2^{\searrow}$, and $N_2^{\downarrow}$ and then combine the results appropriately. However, there is one type of triplet which we miss by doing so, depending on the position of its leaves within the gall containing its lowest common ancestor. This case corresponds to the type-A ambiguous triplets and we count these triplets separately with an extra function $rcount_A$ (see Lemma 5 below). The problem of counting the common resolved triplets between N_1 and a tree is similarly reduced to three instances of counting the common resolved triplets between two phylogenetic trees (covered by Theorem 2) and adjusting the result by using $rcount_A$.

We now present the details. Define $rcount_A(N_1, N_2)$ as the number of resolved triplets $xy|z$ in both N_1 and N_2 such that $\{x, y, z\}$ is a type-A triplet of N_2 with reticulation leaf z. Similarly, define $rcount_A^*(N_1, N_2)$ as the number of resolved triplets $xy|z$ in both N_1 and N_2 such that $\{x, y, z\}$ is a type-A triplet of both N_1 and N_2 with reticulation leaf z. Observe that in general, $rcount_A(N_1, N_2) \neq rcount_A(N_2, N_1)$, but $rcount_A^*(N_1, N_2) = rcount_A^*(N_2, N_1)$ always holds.

The following lemmas express the relationships between $rcount(N_1, N_2)$, $rcount_A(N_1, N_2)$, and $rcount_A^*(N_1, N_2)$.

Lemma 4. *Suppose that $xy|z$ is a resolved triplet such that x, y, z are in the leaf label set. Then $rcount(xy|z, N_2) = rcount(xy|z, N_2^{\swarrow}) + rcount(xy|z, N_2^{\searrow}) - rcount(xy|z, N_2^{\downarrow}) + rcount_A(xy|z, N_2)$.*

Proof. Any resolved triplet $xy|z$ either appears or does not appear in $t(N_2)$. Also, any ambiguous triplet is either of type-A, type-B, or type-C. Hence, we have the following cases.

- (1) $xy|z \notin t(N_2)$.
- (2) $xy|z \in t(N_2)$:
  - (2.1) $xy|z \in t(N_2)$ and $\{x, y, z\}$ is not ambiguous.
  - (2.2) $xy|z \in t(N_2)$ and $\{x, y, z\}$ is a type-A triplet with reticulation leaf z in N_2.
  - (2.3) $xy|z \in t(N_2)$ and $\{x, y, z\}$ is a type-A triplet with reticulation leaf x or y in N_2.
  - (2.4) $xy|z \in t(N_2)$ and $\{x, y, z\}$ is a type-B or type-C triplet.

In case (1), $xy|z \notin t(N_2^{\nearrow}), t(N_2^{\searrow}), t(N_2^{\downarrow})$. Also, $rcount_A(xy|z, N_2) = 0$ since it cannot be a type-A triplet. Hence, $rcount(xy|z, N_2^{\nearrow}) + rcount(xy|z, N_2^{\searrow}) - rcount(xy|z, N_2^{\downarrow}) + rcount_A(xy|z, N_2) = 0$.

In case (2.1), since $xy|z \in t(N_2)$ and $\{x, y, z\}$ is not ambiguous, $xy|z \in t(N_2^{\nearrow}), t(N_2^{\searrow}), t(N_2^{\downarrow})$. Also, $rcount_A(xy|z, N_2) = 0$. Hence, $rcount(xy|z, N_2^{\nearrow}) + rcount(xy|z, N_2^{\searrow}) - rcount(xy|z, N_2^{\downarrow}) + rcount_A(xy|z, N_2) = 1$.

In case (2.2), $xy|z$ appears in either $N_2^{\nearrow}$ or $N_2^{\searrow}$, but not both, and in $N_2^{\downarrow}$. Because we have $rcount_A(xy|z, N_2) = 1$ by definition, $rcount(xy|z, N_2^{\nearrow}) + rcount(xy|z, N_2^{\searrow}) - rcount(xy|z, N_2^{\downarrow}) + rcount_A(xy|z, N_2) = 1$.

Finally, in cases (2.3) and (2.4), $xy|z$ appears in either $N_2^{\nearrow}$ or $N_2^{\searrow}$, but not both, and $xy|z$ does not appear in $N_2^{\downarrow}$. Also, $rcount_A(xy|z, N_2) = 0$ by definition. Thus, $rcount(xy|z, N_2^{\nearrow}) + rcount(xy|z, N_2^{\searrow}) - rcount(xy|z, N_2^{\downarrow}) = 1$.   □

Lemma 5. $rcount(N_1, N_2) = rcount(N_1, N_2^{\nearrow}) + rcount(N_1, N_2^{\searrow}) - rcount(N_1, N_2^{\downarrow}) + rcount_A(N_1, N_2)$.

Proof. Write $rcount(N_1, N_2) = \sum_{xy|z \in N_1} rcount(xy|z, N_2)$. For $xy|z \in t(N_1)$, by Lemma 4, we have $rcount(xy|z, N_2) = rcount(xy|z, N_2^{\nearrow}) + rcount(xy|z, N_2^{\searrow}) - rcount(xy|z, N_2^{\downarrow}) + rcount_A(xy|z, N_2)$.   □

Lemma 6. $rcount_A(N_1, N_2) = rcount_A(N_1^{\nearrow}, N_2) + rcount_A(N_1^{\searrow}, N_2) - rcount_A(N_1^{\downarrow}, N_2) + rcount_A^*(N_1, N_2)$.

Proof. For $xy|z \in t(N_1)$, by an argument identical to the one in the proof of Lemma 4, we have $rcount(N_1, xy|z) = rcount(N_1^{\nearrow}, xy|z) + rcount(N_1^{\searrow}, xy|z) - rcount(N_1^{\downarrow}, xy|z) + rcount_A'(N_1, xy|z)$, where $rcount_A'(N_1, xy|z) = 1$ if $\{x, y, z\}$ is a type-A triplet of N_1 with reticulation leaf z, and $rcount_A'(N_1, xy|z) = 0$ otherwise.

Let $W = \{xy|z : \{x, y, z\}$ is a type-A triplet of N_2 with reticulation leaf $z\}$. Then $rcount_A(N_1, N_2) = \sum_{xy|z \in W} rcount(N_1, xy|z) = \sum_{xy|z \in W} (rcount(N_1^{\nearrow}, xy|z) + rcount(N_1^{\searrow}, xy|z) - rcount(N_1^{\downarrow}, xy|z) + rcount_A'(N_1, xy|z)) = rcount_A(N_1^{\nearrow}, N_2) + rcount_A(N_1^{\searrow}, N_2) - rcount_A(N_1^{\downarrow}, N_2) + rcount_A^*(N_1, N_2)$ □

Next, we discuss the computation of $rcount_A(T_1, N_2)$ and $rcount_A^*(N_1, N_2)$, where N_1 and N_2 are galled trees and T_1 is a phylogenetic tree. (The case of $rcount_A(N_1, T_2)$ where N_1 is a galled tree and T_2 is a tree, needed in Lemma 5,

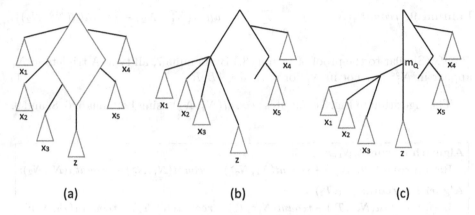

Fig. 2. (a) shows a galled tree N, (b) shows N^L, and (c) shows N^{LL}.

is symmetric.) For $\delta \in \{L, R\}$, denote by τ_δ the tree formed by attaching all subtrees attached to Q_δ to a common root. For any galled tree N, let N^L be a tree formed from $N^\searrow$ by contracting all edges on Q_L for every gall Q (observe that the edges $(r(Q), r(\tau_L))$ and $(r(\tau_L), h_Q)$ are in N^L). Define N^{LL} to be a tree formed from $N^\downarrow$ by replacing all the trees attached to Q_L with τ_L, inserting a new vertex m_Q between $r(Q)$ and h_Q, and replacing the edge $(r(Q), r(\tau_L))$ with $(m_Q, r(\tau_L))$. See Fig. 2 for an example. N^R and N^{RR} are defined analogously.

Lemma 7. *For $\delta \in \{L, R\}$, the following two properties hold.*

- *All resolved triplets in N^δ are in $N^{\delta\delta}$.*
- *All additional resolved triplets $xy|z$ in $N^{\delta\delta}$, i.e., those not in N^δ, are type-A triplets of N with reticulation leaf z.*

Proof. Consider any three leaves $\{x, y, z\}$. If zero, one, or all three of them belong to τ_δ then $N^\delta|\{x, y, z\} = N^{\delta\delta}|\{x, y, z\}$, where $N|W$ is the galled subtree of N formed by retaining only leaves in W. Otherwise, when two of $\{x, y, z\}$ belong to τ_δ, let γ be the lowest common ancestor of x, y, z in $N^{\delta\delta}$. There are two cases:

1. If γ is a proper ancestor of m_Q, then $N^\delta|\{x, y, z\} = N^{\delta\delta}|\{x, y, z\} = xy|z$.
2. Otherwise, $\gamma = m_Q$ and then $N^\delta|\{x, y, z\} = x|y|z$ while $N^{\delta\delta}|\{x, y, z\} = xy|z$.

Hence, all resolved triplets in N^δ are in $N^{\delta\delta}$.

Finally, $xy|z$ is a type-A triplet of N with reticulation leaf z if and only if $\gamma = m_Q$. This shows that the second property also holds. $\qquad\square$

Lemma 8. $rcount_A(T_1, N_2) = \sum_{\delta \in \{L,R\}} \big(rcount(T_1, N_2^{\delta\delta}) - rcount(T_1, N_2^\delta)\big).$

Proof. By Lemma 7, all type-A triplets in N_2 appear in $N_2^{\delta\delta}$ but not in N_2^δ for some $\delta \in \{L, R\}$. The lemma follows. $\qquad\square$

Lemma 9. $rcount_A^*(N_1, N_2) = \sum\limits_{\delta \in \{L,R\}} \left(rcount_A(N_1^{\delta\delta}, N_2) - rcount_A(N_1^{\delta}, N_2) \right).$

Proof. (Similar to the proof of Lemma 8.) By Lemma 7, all type-A triplets in N_1 appear in $N_1^{\delta\delta}$ but not in N_1^{δ} for some $\delta \in \{L, R\}$. $\qquad\square$

The algorithm in Fig. 3 computes $rcount(N_1, N_2)$ using Lemmas 5, 6, 8, and 9.

Algorithm rcount(N_1, N_2)
  Return $rcount(N_1, N_2^{\swarrow}) + rcount(N_1, N_2^{\searrow}) - rcount(N_1, N_2^{\downarrow}) + rcount_A(N_1, N_2)$;

Algorithm rcount(N_1, T_2)
  Return $rcount(N_1^{\swarrow}, T_2) + rcount(N_1^{\searrow}, T_2) - rcount(N_1^{\downarrow}, T_2) + rcount_A(T_2, N_1)$;

Algorithm rcount$_A$(N_1, N_2)
  Return $rcount_A(N_1^{\swarrow}, N_2)$ + $rcount_A(N_1^{\searrow}, N_2)$ − $rcount_A(N_1^{\downarrow}, N_2)$ + $rcount_A^*(N_1, N_2)$;

Algorithm rcount$_A$(T_1, N_2)
  Return $\sum\limits_{\delta \in \{L,R\}} \left(rcount(T_1, N_2^{\delta\delta}) - rcount(T_1, N_2^{\delta}) \right)$;

Algorithm rcount$_A^*$(N_1, N_2)
  Return $\sum\limits_{\delta \in \{L,R\}} \left(rcount_A(N_1^{\delta\delta}, N_2) - rcount_A(N_1^{\delta}, N_2) \right)$;

Fig. 3. The algorithm for computing $rcount(N_1, N_2)$.

Lemma 10. *The algorithm* $rcount(N_1, N_2)$ *in Fig. 3 makes a total of 37 calls to* $rcount(T_1, T_2)$, *where* T_1 *and* T_2 *are phylogenetic trees.*

Proof. First, $rcount_A(T_1, N_2)$ is obtained by making 4 calls to $rcount(T_1, T_2)$, and $rcount_A^*(N_1, N_2)$ by 4 calls to $rcount_A(T_1, T_2)$. Next, $rcount_A(N_1, N_2)$ is obtained by 3 calls to $rcount_A(T_1, N_2)$ and 1 call to $rcount_A^*(N_1, N_2)$. In total, $rcount_A(N_1, N_2)$ uses $3 \cdot 4 + 4 = 16$ calls to $rcount(T_1, T_2)$.

Similarly, $rcount(N_1, T_2)$ is obtained by 3 calls to $rcount(T_1, T_2)$ and 1 call to $rcount_A(T_2, N_1)$. In total, $rcount(N_1, T_2)$ can be computed by $3 \cdot 1 + 4 = 7$ calls to $rcount(T_1, T_2)$.

Finally, $rcount(N_1, N_2)$ is obtained by 3 calls to $rcount(N_1, T_2)$ and 1 call to $rcount_A(N_1, N_2)$. In total, $rcount(N_1, N_2)$ uses $3 \cdot 7 + 16 = 37$ calls to $rcount(T_1, T_2)$. $\qquad\square$

By Theorem 2, $rcount(T_1, T_2)$ can be computed in $O(n \log n)$ time for any two trees T_1, T_2. Lemma 10 shows that the algorithm in Fig. 3 makes a constant number of calls to $rcount(T_1, T_2)$. Lastly, constructing each of the constant number of trees used as arguments to $rcount(T_1, T_2)$ ($N_1^{\swarrow}$, $N_1^{\downarrow}$, etc.) takes $O(n)$ time. Thus, the total running time to obtain $rcount(N_1, N_2)$ is $O(n \log n)$.

3.2   Counting Common Fan Triplets in N_1 and N_2

To compute $fcount(N_1, N_2)$, we modify the technique from the previous subsection. The main difference is that we count type-B and type-C triplets separately. (Some proofs have been omitted from the conference version of the paper.)

Define $fcount_{BC}(N_1, N_2)$ as the number of triplets $\{x, y, z\}$ such that $x|y|z$ is a fan triplet in N_1 and $\{x, y, z\}$ is a type-B or type-C triplet in N_2. Also, define $fcount^*_{BC}(N_1, N_2)$ as the number of type-B and type-C triplets $\{x, y, z\}$ that appear in both N_1 and N_2. Similar to what was done in Sect. 3.1 where $rcount(N_1, N_2)$ was expressed using $rcount_A(N_1, N_2)$ and $rcount^*_A(N_1, N_2)$, we express $fcount(N_1, N_2)$ using $fcount_{BC}(N_1, N_2)$ and $fcount^*_{BC}(N_1, N_2)$.

Lemma 11. *Let $x|y|z$ be a fan triplet. Then $fcount(x|y|z, N_2) = fcount(x|y|z, N_2^{\swarrow}) + fcount(x|y|z, N_2^{\searrow}) - fcount(x|y|z, N_2^{\downarrow}) + fcount_{BC}(x|y|z, N_2) = 0$.*

Lemma 12. *$fcount(N_1, N_2) = fcount(N_1, N_2^{\swarrow}) + fcount(N_1, N_2^{\searrow}) - fcount(N_1, N_2^{\downarrow}) + fcount_{BC}(N_1, N_2)$.*

Lemma 13. *$fcount_{BC}(N_1, N_2) = fcount_{BC}(N_1^{\swarrow}, N_2) + fcount_{BC}(N_1^{\searrow}, N_2) - fcount_{BC}(N_1, N_2^{\downarrow}) + fcount^*_{BC}(N_1, N_2)$.*

The rest of this subsection considers how to compute $fcount_{BC}(T_1, N_2)$ and $fcount^*_{BC}(N_1, N_2)$ efficiently. A *caterpillar* is a binary phylogenetic tree in which every internal node has at least one leaf child. Given a galled tree N, we define N^B as the tree formed by performing the following three steps on a copy of N:

- Replace every degree-k vertex which is not a split vertex in N by a length-k caterpillar.
- For every gall Q, if the split vertex $r(Q)$ is of degree $k > 2$, for all $k-2$ children of $r(Q)$ which are not on the gall, replace them by a length-$(k-2)$ caterpillar and attach it $r(Q)$. Furthermore, creating a new vertex u and attach the roots of Q_L and Q_R to u and attach u to $r(Q)$.
- Remove the reticulation vertex h_Q's two parent edges and attach h_Q and its subtrees to $r(Q)$.

Also, we define N^C as a tree obtained by performing the following three steps on a copy of N:

- Replace every degree-k vertex which is not a split vertex in N by a length-k caterpillar.
- For every gall Q, if the split vertex $r(Q)$ is of degree $k > 2$, for all $k - 2$ children of $r(Q)$ which are not on the gall, replace them by a length-$(k - 2)$ caterpillar and attach it to a new vertex between $r(Q)$ and its parent.
- Remove the reticulation vertex h_Q's two parent edges and attach h_Q and its subtrees to $r(Q)$.

Figure 4 gives an example illustrating how to construct N^B and N^C from N. The next lemma states how N, N^B, and N^C are related.

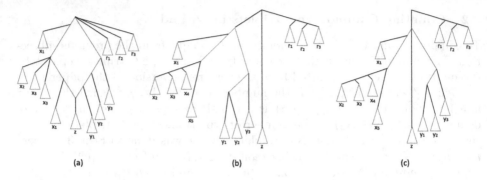

Fig. 4. (a) is an example of a galled tree N, (b) is N^B, and (c) is N^C.

Lemma 14. *(1) $\{x, y, z\}$ is a type-B triplet in N if and only if $x|y|z$ is a fan triplet in N^B; (2) $\{x, y, z\}$ is a type-C triplet in N if and only if $x|y|z$ is a fan triplet in N^C.*

Proof. By construction, every vertex in N^B and N^C has 2 or 3 children. Any vertex in N^B and N^C with 3 children corresponds to a split vertex of a gall in N.

For (1): ($\rightarrow$) If $\{x, y, z\}$ is a type-B triplet in N, there exists a gall Q in N such that x, y, z are in three different subtrees attached to Q where one leaf (say, x) is in a subtree attached to Q_L or Q_R, another leaf (say, y) is in a subtree attached to $r(Q)$ and the remaining leaf (say, z) is in a subtree attached to h_Q. Then, in N^B, by construction, $x|y|z$ is a fan triplet in N^B.

($\leftarrow$) If $x|y|z$ is a fan triplet in N^B, let u be the lowest common ancestor of x, y, z in N^B. u is of degree-3 and it corresponds to a gall Q. This implies, x, y, z are in a subtree attached to $r(Q)$, a subtree attached to h_Q and a subtree attached to Q_L or Q_R. Hence, $\{x, y, z\}$ is a type-B triplet in N.

For (2): ($\rightarrow$) If $\{x, y, z\}$ is a type-C triplet in N, there exists a gall Q in N such that x, y, z are in three different subtrees attached to Q. Where one leaf (say, x) is in a subtree attached to Q_L, another leaf (say, y) is in a subtree attached to Q_R, and the remaining leaf (say, z) is in a subtree attached to h_Q. Then, in N^C, by construction, $x|y|z$ is a fan triplet in N^B.

($\leftarrow$) If $x|y|z$ is a fan triplet in N^B, let u be the lowest common ancestor of x, y, z in N^B. u is of degree-3 and it corresponds to a gall Q. This implies, x, y, z are in a subtree attached to Q_L, a subtree attached to Q_R and a subtree attached to h_Q. Hence, $\{x, y, z\}$ is a type-C triplet in N.   □

We have the following two lemmas.

Lemma 15. $fcount_{BC}(T_1, N_2) = fcount(T_1, N_2^B) + fcount(T_1, N_2^C)$.

Lemma 16. $fcount^*_{BC}(N_1, N_2) = fcount(N_1^B, N_2^B) + fcount(N_1^C, N_2^C)$.

The algorithm in Fig. 5 computes $fcount(N_1, N_2)$ by combining Lemmas 12, 13, 15, and 16.

Algorithm fcount(N_1, N_2)
  Return $fcount(N_1, N_2^{\nearrow}) + fcount(N_1, N_2^{\searrow}) - fcount(N_1, N_2^{\downarrow}) + fcount_{BC}(N_1, N_2)$;

Algorithm fcount(N_1, T_2)
  Return $fcount(N_1^{\nearrow}, T_2) + fcount(N_1^{\searrow}, T_2) - fcount(N_1, T_2^{\downarrow}) + fcount_{BC}(T_2, N_1)$;

Algorithm fcount$_{BC}(N_1, N_2)$
  Return $fcount_{BC}(N_1^{\nearrow}, N_2) + fcount_{BC}(N_1^{\searrow}, N_2) - fcount_{BC}(N_1, N_2^{\downarrow}) + fcount_{BC}^*(N_1, N_2)$;

Algorithm fcount$_{BC}(T_1, N_2)$
  Return $fcount(T_1, N_2^B) + fcount(T_1, N_2^C)$;

Algorithm fcount$_{BC}^*(N_1, N_2)$
  Return $fcount(N_1^B, N_2^B) + fcount(N_1^C, N_2^C)$;

Fig. 5. The algorithm for computing $fcount(N_1, N_2)$.

Lemma 17. *The algorithm $fcount(N_1, N_2)$ in Fig. 5 makes a total of 23 calls to $fcount(T_1, T_2)$, where T_1 and T_2 are phylogenetic trees.*

Proof. Note that $fcount_{BC}(T_1, N_2)$ is computed by 2 calls to $fcount(T_1, T_2)$, and $fcount_{BC}^*(N_1, N_2)$ by 2 calls to $fcount_A(T_1, T_2)$. Moreover, $fcount_{BC}(N_1, N_2)$ makes 3 calls to $fcount_{BC}(T_1, N_2)$ and 1 call to $fcount_{BC}^*(N_1, N_2)$. In total, $fcount_{BC}(N_1, N_2)$ uses $3 \cdot 2 + 2 = 8$ calls to $fcount(T_1, T_2)$.

In the same way, $fcount(N_1, T_2)$ makes 3 calls to $fcount(T_1, T_2)$ and 1 call to $fcount_{BC}(T_2, N_1)$. In total, $fcount(N_1, T_2)$ is obtained from $3 \cdot 1 + 2 = 5$ calls to $fcount(T_1, T_2)$.

Finally, $fcount(N_1, N_2)$ is obtained from 3 calls to $fcount(N_1, T_2)$ and 1 call to $rcount_{BC}(N_1, N_2)$. In total, $fcount(N_1, N_2)$ makes $3 \cdot 5 + 8 = 23$ calls to $fcount(T_1, T_2)$. ☐

Since $fcount(T_1, T_2)$ can be computed in $O(n \log n)$ time for any two trees T_1, T_2 by Theorem 2, $fcount(T_1, T_2)$ is called a constant number of times according to Lemma 17, and constructing each of the constant number of trees used as arguments to $fcount(T_1, T_2)$ takes $O(n)$ time, the algorithm in Fig. 5 runs in $O(n \log n)$ time.

4 Concluding Remarks

The presented algorithm requires a subroutine for computing the rooted triplet distance between two phylogenetic trees. If a faster algorithm for the case of trees than the one referred to in Theorem 2 is discovered (e.g., running in $O(n \log \log n)$ time), this would immediately imply a faster algorithm for the case of galled trees as well. As an alternative, the algorithm in [10] was shown experimentally to be faster for reasonably sized inputs; hence its usage may be preferred in a practical setting.

Possible future work is to implement the new algorithm and evaluate its performance in practice. Although the number of calls to the subroutine for computing the rooted triplet distance between trees is constant, the constant is quite large $(37 + 23 = 60)$. Is it possible to reduce this number? An implementation of the new algorithm would benefit considerably by doing so.

An open problem is to determine whether the techniques used here can be extended to compute the rooted triplet distance between more general phylogenetic networks than galled trees.

Acknowledgments. J.J. was partially funded by The Hakubi Project at Kyoto University and KAKENHI grant number 26330014.

References

1. Bansal, M.S., Dong, J., Fernández-Baca, D.: Comparing and aggregating partially resolved trees. Theor. Comput. Sci. **412**(48), 6634–6652 (2011)
2. Brodal, G.S., Fagerberg, R., Mailund, T., Pedersen, C.N.S., Sand, A.: Efficient algorithms for computing the triplet and quartet distance between trees of arbitrary degree. In: Proceedings of the 24th Annual ACM-SIAM Symposium on Discrete Algorithms (SODA 2013), pp. 1814–1832. SIAM (2013)
3. Cardona, G., Llabrés, M., Rosselló, R., Valiente, G.: Comparison of galled trees. IEEE/ACM Trans. Comput. Biol. Bioinf. **8**(2), 410–427 (2011)
4. Critchlow, D.E., Pearl, D.K., Qian, C.: The triples distance for rooted bifurcating phylogenetic trees. Syst. Biol. **45**(3), 323–334 (1996)
5. Dobson, A.J.: Comparing the shapes of trees. In: Street, A.P., Wallis, W.D. (eds.) Combinatorial Mathematics III. LNM, vol. 452, pp. 95–100. Springer, Heidelberg (1975). doi:10.1007/BFb0069548
6. Gambette, P., Huber, K.T.: On encodings of phylogenetic networks of bounded level. J. Math. Biol. **65**(1), 157–180 (2012)
7. Gusfield, D., Eddhu, S., Langley, C.: Optimal, efficient reconstruction of phylogenetic networks with constrained recombination. J. Bioinf. Comput. Biol. **2**(1), 173–213 (2004)
8. Huson, D.H., Rupp, R., Scornavacca, C.: Phylogenetic Networks: Concepts Algorithms and Applications. Cambridge University Press, Cambridge (2010)
9. Jansson, J., Lingas, A.: Computing the rooted triplet distance between galled trees by counting triangles. J. Discrete Algorithms **25**, 66–78 (2014)
10. Jansson, J., Rajaby, R.: A more practical algorithm for the rooted triplet distance. J. Comput. Biol. **24**(2), 106–126 (2017)
11. Kuhner, M.K., Felsenstein, J.: A simulation comparison of phylogeny algorithms under equal and unequal evolutionary rates. Mol. Biol. Evol. **11**(3), 459–468 (1994)
12. Morrison, D.: Introduction to Phylogenetic Networks. RJR Productions (2011)
13. Wang, L., Ma, B., Li, M.: Fixed topology alignment with recombination. Discrete Appl. Math. **104**(1–3), 281–300 (2000)

Clustering the Space of Maximum Parsimony Reconciliations in the Duplication-Transfer-Loss Model

Alex Ozdemir[1], Michael Sheely[1], Daniel Bork[2], Ricson Cheng[2], Reyna Hulett[3],
Jean Sung[1], Jincheng Wang[1], and Ran Libeskind-Hadas[1(✉)]

[1] Department of Computer Science, Harvey Mudd College, 301 Platt Blvd.,
Claremont, CA 91711, USA
{aozdemir,msheely,jsung,jwang,hadas}@g.hmc.edu
[2] Department of Computer Science, Carnegie Mellon University,
5000 Forbes Ave, Pittsburgh, PA 15213, USA
{dbork,rcheng}@andrew.cmu.edu
[3] Department of Computer Science, Stanford University,
450 Serra Mall, Stanford, CA 94305, USA
reyna.hulett@cs.stanford.edu

Abstract. Phylogenetic tree reconciliation is widely used in the fields of molecular evolution, cophylogenetics, parasitology, and biogeography for studying the evolutionary histories of pairs of entities. Reconciliation is often performed using maximum parsimony under the DTL (Duplication-Transfer-Loss) event model. Since the number of maximum parsimony reconciliations (MPRs) can be exponential in the sizes of the trees, an important problem is that of finding a small number of representative reconciliations. We give a polynomial time algorithm that can be used to find the cluster representatives of the space of MPRs with respect to a number of different clustering algorithms and specified number of clusters.

Keywords: Tree reconciliation · Duplication-Transfer-Loss model · Clustering

1 Introduction

Phylogenetic tree reconciliation is an important technique for studying the evolutionary histories of pairs of entities such as gene families and species, parasites and their hosts, and species and their geographical habitats. The reconciliation problem takes as input two trees and the associations between their leaves and seeks to find a mapping between the trees that accounts for their topological incongruence with respect to a given set of biological events. In the widely-used DTL model the four event types are *speciation*, *duplication*, *transfer*, and

This work was funded by the U.S. National Science Foundation under Grant Numbers IIS-1419739 and 1433220.

© Springer International Publishing AG 2017
D. Figueiredo et al. (Eds.): AlCoB 2017, LNBI 10252, pp. 127–139, 2017.
DOI: 10.1007/978-3-319-58163-7_9

loss [1–6, 16]. We denote the two trees as the *species tree* (S) and the *gene tree* (G), although these trees could be host and species trees or area cladograms and species trees in the contexts of cophylogenetic and biogeographical studies, respectively.

Reconciliation in the DTL model is typically performed using a maximum parsimony formulation, where each event type has an assigned cost and the objective is to find a reconciliation of minimum total cost, called a *maximum parsimony reconciliation* or *MPR*. Efficient algorithms are known for finding MPRs in the DTL model [1, 15].

In general, the number of MPRs can grow exponentially with the sizes of the species and gene trees [13]. Consequently, a number of efforts have been made to summarize the vast space of MPRs. Nguyen *et al.* [10] showed that choosing a single random MPR can lead to inaccurate inferences and gave an efficient algorithm to compute a median MPR. Median MPRs were subsequently used to summarize reconciliation space in [14]. Bansal *et al.* [2] showed how a sample of MPRs can be selected uniformly at random. Ma *et al.* [9] examined the problem of finding a set of k reconciliations that collectively cover the most frequently occurring events in MPR space, for a given number k.

We study the problem of clustering the space of MPRs, both to represent the space by a small number of cluster representatives and to gain insights into the structure of the space. We first define the *reconciliation count function* that can be used to implement a number of different clustering algorithms for MPR space. We then show that the reconciliation count function can be computed exactly in polynomial time (where the degree of the polynomial depends on the number of clusters), in spite of the fact that the number of MPRs can be exponential in the size of the given trees. Our results leverage the seminal work of Scornavacca *et al.* [12] and Nguyen *et al.* [10].

We demonstrate the utility of the reconciliation count function by showing how it can be used to implement two clustering algorithms for k-*medoids* and k-*centers*. The k-*medoids* problem seeks to find a representative set of k MPRs, called *medoids*, such that the sum of the distances between each MPR and its nearest medoid is minimized. Similarly, k-*centers* seeks to identify a representative set of k MPRs, known as *centers*, such that the maximum distance between each MPR and its nearest center is minimized. These are just two examples of the many clustering algorithms that can be implemented using reconciliation counts.

This work provides new tools for exploring the structure and diversity of MPR space that may not be gleaned from a single median reconciliation or a set of randomly sampled reconciliations. Thus, the results presented here are potentially useful to practitioners who wish to better understand the space of optimal reconciliations for specific data sets as well as for researchers seeking to gain better insights into MPR space in general.

In summary, in this paper:

1. We define the *reconciliation count function* and give a polynomial-time algorithm for computing it.

2. We demonstrate the utility of the reconciliation count function by showing how it can be used to solve the k-medoids and k-centers problems for MPR space and to compute statistics on the clusterings.
3. In the Supplementary Materials, we give experimental results using the Tree of Life data set [5], comparing the k-medoids and k-centers representatives to randomly selected ones. (See www.cs.hmc.edu/~hadas/supplement.pdf).
4. We provide an implementation of our algorithms. (See www.cs.hmc.edu/~hadas/clusters.zip).

2   Definitions

In this section we give definitions and notation used throughout this paper. In the interest of brevity, we provide the minimum background required to develop our results in the subsequent sections. For completeness, formal definitions are given in the Supplementary Materials.

2.1   Maximum Parsimony Reconciliations

An instance of the DTL maximum parsimony reconciliation problem comprises a gene tree G, a species tree S, a leaf mapping Le from the leaves of G to the leaves of S (which need not be one-to-one nor onto), and positive costs for duplication, transfer, and loss events. We assume that the trees are undated in the sense that no information is given about the relative times of speciation events in either the gene or species trees. A reconciliation is a mapping $\mathcal{M}$ of the vertices of G into the vertices of S that is consistent with the leaf mapping Le and, for each internal gene vertex g with children g' and g'', neither $\mathcal{M}(g')$ nor $\mathcal{M}(g'')$ are ancestors of $\mathcal{M}(g)$ and at least one of $\mathcal{M}(g')$ and $\mathcal{M}(g'')$ is either equal to, or a descendant of, $\mathcal{M}(g)$.

The mapping $\mathcal{M}$ induces speciation, duplication, transfer, and loss events. While the formal definitions of these events are given in the Supplementary Materials, the following suffices for our treatment: Let g be an internal vertex in G with children g' and g''. Vertex g is a speciation vertex if $\mathcal{M}(g)$ is the most recent common ancestor of $\mathcal{M}(g')$ and $\mathcal{M}(g'')$ and $\mathcal{M}(g')$ and $\mathcal{M}(g'')$ are neither equal to one another nor ancestrally related. Vertex g is a duplication vertex if $\mathcal{M}(g)$ is the most recent common ancestor of $\mathcal{M}(g')$ and $\mathcal{M}(g'')$ but $\mathcal{M}(g')$ and $\mathcal{M}(g'')$ are either equal or ancestrally related. A duplication can be viewed as a mapping of gene vertex g onto the edge from the parent of $\mathcal{M}(g)$ to $\mathcal{M}(g)$. Vertex g is a transfer vertex if exactly one of $\mathcal{M}(g')$ or $\mathcal{M}(g'')$ is a descendant of $\mathcal{M}(g)$ and the other is not in the subtree rooted at $\mathcal{M}(g)$. A loss event arises for each internal vertex on the path in S from $\mathcal{M}(g)$ to $\mathcal{M}(g')$, for each parent-child pair (g, g').[1]

---

[1] This characterization slightly simplifies the way that losses are actually counted and omits details about losses arising from transfer events. While this suffices for presenting our work, full details are given in the Supplementary Materials.

The objective of the DTL maximum parsimony reconciliation problem is to find a reconciliation that minimizes the sum of the number of duplication, transfer, and loss events weighted by their respective event costs. We henceforth refer to maximum parsimony reconciliations as *MPRs*. A number of similar dynamic programming algorithms have been given for finding MPRs in time $O(|G||S|)$ [1,16]. Figure 1(a) shows a species tree, gene tree, and a leaf mapping and (b) and (c) show two different MPRs.

2.2   Reconciliation Graphs

Scornavacca *et al.* [12] developed a data structure called a *reconciliation graph* for compactly representing the space of all MPRs for dated trees. Ma *et al.* [9] adopted reconciliation graphs for undated trees. For the purposes of the remainder of this paper, the following characterization suffices (with full details in the Supplementary Materials):

The reconciliation graph for an instance of the DTL MPR problem (comprising trees G, S, leaf mapping Le and given DTL event costs) is a directed acyclic graph (DAG) that consists of *mapping vertices* and *event vertices* and directed edges between these two vertex types. Specifically, the graph contains a *mapping vertex* for each (g, s) pair such that $\mathcal{M}(g) = s$ for some MPR $\mathcal{M}$ and an *event vertex* for each event in which $\mathcal{M}(g) = s$, with a directed edge from mapping vertex (g, s) to each such event vertex. (Note that a mapping vertex (g, s) may have edges to multiple event vertices since, for example, g may be mapped to s as a speciation event in one MPR and as a transfer event in a different MPR.) Let g' and g'' denote the children of g. If some MPR, $\mathcal{M}$, contains an event in which $\mathcal{M}(g) = s$, $\mathcal{M}(g') = s'$, $\mathcal{M}(g'') = s''$ then that event vertex for (g, s) has a directed edge to mapping vertices (g', s') and (g'', s''). Thus, each speciation, duplication, and transfer event vertex has out-degree 2. Each loss event is represented by an event vertex with out-degree 1 corresponding to a loss induced by a particular vertex in the species tree. Finally, each leaf association $(g, Le(g))$ has a corresponding event vertex which is a sink (vertex of out-degree 0) of the reconciliation graph. The reconciliation graph can be constructed in time $O(|G||S|^2)$ [9]. The right side of Fig. 1 shows the reconciliation graph for the problem instance in Fig. 1(a) with DTL costs 1, 4, and 1 respectively.

This brings us to the two key results that we need in the remainder of this paper. First, there is a bijection between MPRs and subgraphs of the reconciliation graph called *reconciliation trees*. A reconciliation tree begins with a mapping vertex of the form (g, s) where g is the root of the gene tree (but s is not necessarily the root of the species tree since, as shown in Fig. 1(b), a reconciliation need not involve the root of S). The mapping vertex (g, s) is followed by a directed edge to any one neighbor in the reconciliation graph, which is an event vertex corresponding to an event in which g is mapped to s. Next, both neighbors of that event vertex are included; each is a mapping vertex corresponding to the mapping of a child of g. From each such mapping vertex, we again choose any single event vertex neighbor. This process (formalized in the Supplementary Materials) is repeated

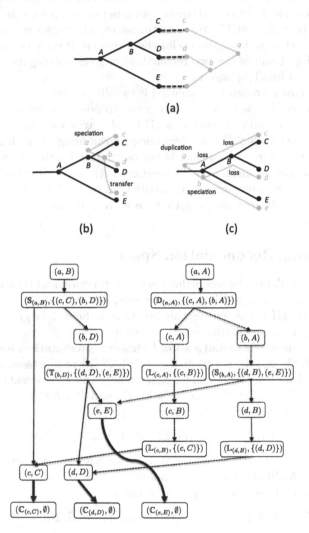

Fig. 1. Top: (a) An instance of the DTL reconciliation problem comprising species tree (black), gene tree (gray), and leaf mapping. Duplication, transfer and loss costs are 1, 4, and 1, respectively. (b) A reconciliation with one speciation and one transfer. (c) A reconciliation with one speciation, one duplication, and three losses. Both reconciliations are MPRs with total cost 4. **Bottom:** The reconciliation graph for the DTL instance in (a). Vertices with solid boundaries are event vertices and those with dashed boundaries are mapping vertices. Event vertices are designated with $\mathbb{S}$ (speciation event), $\mathbb{D}$ (duplication event), $\mathbb{T}$ (transfer event), $\mathbb{L}$ (loss event), and $\mathbb{C}$ (leaf association). The reconciliation tree indicated by solid edges corresponds to the MPR in (b) and the reconciliation tree indicated by dashed edges corresponds to the MPR in (c), with bold edges representing the shared parts of the two reconciliations.

until the sinks of the reconciliation graph are reached. It is not difficult to show that this process yields a tree and the bijection between these reconciliations trees and MPRs is proved in [9,12]. Henceforth, we use the terms *reconciliation trees* and MPRs interchangeably as we do for the terms *event vertices* and *events*. The right side of Fig. 1 shows the two reconciliation trees corresponding to the two reconciliations in Fig. 1(b) and (c).

The second major result that we need is as follows: Given a *score* function, σ, that maps vertices in the reconciliation graph to non-negative real numbers, we can find a reconciliation tree (that is, a MPR) of *maximum* total score in polynomial via a simple $O(|G||S|^2)$ time dynamic programming algorithm [9,10]. Note that this *maximization* problem should not be confused with the problem of finding a *minimum* cost reconciliation. The vertices in the reconciliation graph *a priori* represent events and mappings in minimum cost reconciliations. The score $\sigma(v)$ can represent an arbitrary quantity that we wish to maximize over all minimum cost reconciliations.

3   Clustering Reconciliation Space

In this section we define the reconciliation count function and then show how two well-known clustering algorithms, one for medoids and one for centers, can be implemented for MPR space using this function. In Sect. 4 we give the algorithm for computing the reconciliation count function.

In general, our goal is to find a set of k cluster representatives for a given clustering method. We use the notation $\mathcal{T} = \{T_1, \ldots, T_k\}$ to represent such a set of k MPRs. Let R denote an MPR and let $\mathbb{E}(R)$ denote the set of events in R. For any two MPRs R_1, R_2, let the distance between R_1 and R_2, $d(R_1, R_2)$, be the size of the symmetric set difference of the event sets [10]:[2]

$$d(R_1, R_2) = |\mathbb{E}(R_1) \backslash \mathbb{E}(R_2)| + |\mathbb{E}(R_2) \backslash \mathbb{E}(R_1)|$$

For convenience, we define the distance between a reconciliation and a set $\mathcal{T} = \{T_1, \ldots, T_k\}$ of MPRs as a k-dimensional vector that contains the distance from the reconciliation to each reconciliation in the set:

$$d(R, \mathcal{T}) = d(R, \{T_1, T_2, \ldots, T_k\}) = [d(R, T_1), d(R, T_2), \ldots, d(R, T_k)]$$

Definition 1 (Reconciliation Count Function). *Let $\mathcal{R}$ denote the set of all MPRs for a given DTL problem instance and let $\mathcal{T} = \{T_1, \ldots, T_k\}$ denote a set of k MPRs in $\mathcal{R}$. Let v denote an event vertex in the reconciliation graph. The reconciliation count function $Count_{\mathcal{T},v} : \mathbb{N}^k \to \mathbb{N}$ maps $\mathbf{d} \in \mathbb{N}^k$ to the number of reconciliations $R \in \mathcal{R}$ that contain event vertex v and $d(R, \mathcal{T}) = \mathbf{d}$. Let $Count_{\mathcal{T}}(\mathbf{d})$ denote the number of reconciliations $R \in \mathcal{R}$ with $d(R, \mathcal{T}) = \mathbf{d}$, regardless of whether or not they contain a particular event vertex.*

In the next two sections we demonstrate the utility of the reconciliation count function by showing how it can be used to solve the k-medoids and k-centers problems for MPR space.

---

[2] Other distance functions are also possible.

3.1 k-medoids

Let $\mathscr{R}$ denote the set of all MPRs for a given instance of the DTL reconciliation problem. Let k be a positive integer. A set of k MPRs, $\mathcal{T} = \{T_1, T_2, ...T_k\} \subseteq \mathscr{R}$, induces a partition of $\mathscr{R}$ into clusters $C_{T_1}, C_{T_2}, ..., C_{T_k}$ such that C_{T_i} denotes the set of MPRs that are closer to T_i than to any other MPR in $\mathcal{T}$, breaking ties arbitrarily. The k-medoids problem seeks to find a set $\mathcal{T}$ of k MPRs that minimizes the sum of the distances between each MPR in $\mathscr{R}$ and a closest MPR in $\mathcal{T}$. More formally, the objective is to find:

$$\underset{\substack{\mathcal{T} \subset \mathscr{R} \\ |\mathcal{T}|=k}}{\arg\min} \sum_{T_i \in \mathcal{T}} \sum_{R \in C_{T_i}} d(R, T_i)$$

The elements of $\mathcal{T}$ are called *medoids*. Since the k-medoids problem is NP-hard in general [7], heuristics are used to find good, but not necessarily optimal, solutions. Park *et al.* [11] offer a simple and effective local search algorithm, with $\mathcal{T}$ initially set to k arbitrarily chosen points (MPRs in our case). In each iteration, the approximate medoid of each cluster is replaced with the MPR in that cluster that minimizes the sum of distances within that cluster. In practice, the algorithm iterates until some termination condition is reached (e.g., a maximum number of iterations). The algorithm is given below.

Algorithm 1. k-medoids heuristic [11]

1: **procedure** K-MEDOIDS($\mathscr{R}, k$)
2:     $\mathcal{T} \leftarrow k$ arbitrary MPRs in $\mathscr{R}$
3:     **while** some termination condition is not satisfied **do**
4:         **for** $T_i \in \mathcal{T}$ **do**
5:             $T_i \leftarrow \arg\min_{M \in \mathscr{R}} \sum_{R \in C_{T_i}} d(R, M)$

6:     **return** $\mathcal{T}$

Since the number of MPRs can be exponentially large, we cannot compute line 5 in polynomial time by evaluating $\sum_{R \in C_{T_i}} d(R, M)$ for each M.

To address this problem, we define the set of functions $\{g_i : \mathbb{N}^k \to \{0,1\}\}$, each of which takes a vector of the distances from an MPR $R \in \mathscr{R}$ to each in $T_j \in \mathcal{T}$ as an input and determines whether R is in C_{T_i}. That is,

$$g_i(\mathbf{d} = [d_1, d_2, \ldots, d_k]) = \begin{cases} 1 & \text{if } i \text{ is the least index such that } d_i \leq d_j \text{ for all } 1 \leq j \leq k \\ 0 & \text{otherwise} \end{cases}$$

Now, we can rewrite line 5 in the algorithm as

$$\underset{M \in \mathscr{R}}{\arg\min} \sum_{R \in \mathscr{R}} d(R, M) \cdot g_i(d(R, \mathcal{T})) \tag{1}$$

By definition, $d(R, M) = |\mathbb{E}(R) \backslash \mathbb{E}(M)| + |\mathbb{E}(M) \backslash \mathbb{E}(R)|$. Moreover:

$$|\mathbb{E}(R) \backslash \mathbb{E}(M)| + |\mathbb{E}(M) \backslash \mathbb{E}(R)| = |\mathbb{E}(R) \cup \mathbb{E}(M)| - |\mathbb{E}(R) \cap \mathbb{E}(M)|$$
$$= |\mathbb{E}(R)| + |\mathbb{E}(M)| - 2|\mathbb{E}(R) \cap \mathbb{E}(M)|$$

Thus, we can rewrite (1) as:

$$\underset{M \in \mathscr{R}}{\arg \min} \sum_{R \in \mathscr{R}} \left(|\mathbb{E}(R)| - 2|\mathbb{E}(M) \cap \mathbb{E}(R)| + |\mathbb{E}(M)| \right) \cdot g_i \left(d(R, \mathcal{T}) \right) \qquad (2)$$

Since $|\mathbb{E}(R)|$ does not depend on M, the minimization problem in (2) is equivalent to the following maximization problem:

$$\underset{M \in \mathscr{R}}{\arg \max} \sum_{R \in \mathscr{R}} \left(2|\mathbb{E}(M) \cap \mathbb{E}(R)| - |\mathbb{E}(M)| \right) \cdot g_i \left(d(R, \mathcal{T}) \right) \qquad (3)$$

Next, we rewrite this as a summation over the events in M:

$$\underset{M \in \mathscr{R}}{\arg \max} \sum_{e \in \mathbb{E}(M)} \sum_{R \in \mathscr{R}} \left(2 \, |\{e\} \cap \mathbb{E}(R)| - 1 \right) \cdot g_i \left(d(R, \mathcal{T}) \right) \qquad (4)$$

We then split the sum to yield:

$$\underset{M \in \mathscr{R}}{\arg \max} \sum_{e \in \mathbb{E}(M)} \left(\sum_{R \in \mathscr{R}} 2 \, |\{e\} \cap \mathbb{E}(R)| \cdot g_i \left(d(R, \mathcal{T}) \right) - \sum_{R \in \mathscr{R}} g_i \left(d(R, \mathcal{T}) \right) \right) \qquad (5)$$

Define $S(e)$ to be the set of all reconciliations containing event e. We rewrite the first inner summation as a sum over $S(e)$, since $|\{e\} \cap \mathbb{E}(R)|$ is 1 for all $R \in S(e)$ and 0 for all $R \notin S(e)$:

$$\underset{M \in \mathscr{R}}{\arg \max} \sum_{e \in \mathbb{E}(M)} \left(\sum_{R \in S(e)} 2 \cdot g_i \left(d(R, \mathcal{T}) \right) - \sum_{R \in \mathscr{R}} g_i \left(d(R, \mathcal{T}) \right) \right) \qquad (6)$$

Define $f(\mathbf{d}, X)$ to be the set of reconciliations in X such that $d(R, \mathcal{T}) = \mathbf{d}$. Notice that we can partition $S(e)$ as $\{f(\mathbf{d}, S(e)) \mid \mathbf{d} \in \mathbb{N}^k\}$ and $\mathscr{R}$ as $\{f(\mathbf{d}, \mathscr{R}) \mid \mathbf{d} \in \mathbb{N}^k\}$. Then we can rewrite our sum over these partitions as:

$$\underset{M \in \mathscr{R}}{\arg \max} \sum_{e \in \mathbb{E}(M)} \sum_{\mathbf{d} \in \mathbb{N}^k} \left(\sum_{R \in f(\mathbf{d}, S(e))} 2 \cdot g_i \left(d(R, \mathcal{T}) \right) - \sum_{R \in f(\mathbf{d}, \mathscr{R})} g_i \left(d(R, \mathcal{T}) \right) \right) \qquad (7)$$

The inner terms can now be simplified, yielding:

$$\underset{M \in \mathscr{R}}{\arg \max} \sum_{e \in \mathbb{E}(M)} \sum_{\mathbf{d} \in \mathbb{N}^k} \left(\sum_{R \in f(\mathbf{d}, S(e))} 2 \cdot g_i(\mathbf{d}) - \sum_{R \in f(\mathbf{d}, \mathscr{R})} g_i(\mathbf{d}) \right) \qquad (8)$$

Now we define $\sigma : V(\mathcal{G}) \rightarrow \mathbb{R}$ as a function from the vertices of the reconciliation graph to the reals as follows:

$$\sigma(v) = \begin{cases} \sum_{\mathbf{d} \in \mathbb{N}^k} \left(2 \cdot |f(\mathbf{d}, S(v))| - |f(\mathbf{d}, \mathcal{R})|\right) \cdot g_i(\mathbf{d}) & \text{if } v \text{ is an event vertex} \\ 0 & \text{if } v \text{ is a mapping vertex} \end{cases}$$

Notice that $|f(\mathbf{d}, S(v))| = \text{Count}_{\mathcal{T},v}(\mathbf{d})$ and $|f(\mathbf{d}, \mathcal{R})| = \text{Count}_{\mathcal{T}}(\mathbf{d})$ so we have:

$$\sigma(v) = \begin{cases} \sum_{\mathbf{d} \in \mathbb{N}^k} \left(2 \cdot \text{Count}_{\mathcal{T},v}(\mathbf{d}) - \text{Count}_{\mathcal{T}}(\mathbf{d})\right) \cdot g_i(\mathbf{d}) & \text{if } v \text{ is an event vertex} \\ 0 & \text{if } v \text{ is a mapping vertex} \end{cases}$$

This reduces to finding:

$$\underset{M \in \mathcal{R}}{\arg\max} \sum_{v \in V(M)} \sigma(v) \tag{9}$$

As noted in the previous section, the problem of finding a reconciliation that maximizes the score of its constituent event vertices can be solved in $O(|G||S|^2)$ time by dynamic programming [9,10]. Thus, we have shown that the widely-used k-medoids heuristic of Park et $al.$ [11] can be applied to MPR space by computing the reconciliation count function $\text{Count}_{\mathcal{T},v}$.

3.2 k-centers

The objective of the k-centers problem is to find a set $\mathcal{T}$ of k MPRs that minimizes the $covering$ $radius$, the maximum distance between any MPR and the nearest element in $\mathcal{T}$. More formally, letting MC denote the function that returns the minimum component of a vector, we seek to find:

$$\underset{\substack{\mathcal{T} \subset \mathcal{R} \\ |\mathcal{T}|=k}}{\arg\min} \underset{R \in \mathcal{R}}{\max} MC(d(R, \mathcal{T}))$$

While this problem is NP-complete [7], Gonzalez [8] proved that the following algorithm finds solutions that are guaranteed to be within a factor of two of optimal:

Algorithm 2. k-centers 2-approximation [8]

1: **procedure** K-CENTERS($\mathcal{R}, k$)
2:     $\mathcal{T} \leftarrow \{\text{arbitrary } R \text{ in } \mathcal{R}\}$
3:     **for** $k - 1$ iterations **do**
4:         $T \leftarrow \arg\max_{R \in \mathcal{R}} MC(d(R, \mathcal{T}))$
5:         $\mathcal{T} \leftarrow \mathcal{T} \cup \{T\}$
6:     **return** $\mathcal{T}$

Line 4 cannot be efficiently computed by simple iteration because there can be exponentially many MPRs. Again, the reconciliation count function will be applied to perform this step in polynomial time.

Given some number of current centers, line 4 of Algorithm 2 seeks a reconciliation that maximizes the minimum distance from the rest. Given the reconciliation counts to the current centers, we can find this distance by the following algorithm.

Algorithm 3. Farthest from centers

```
1: procedure FFC(T)
2:     d* ← 0
3:     for d ∈ N^k do
4:         if Count_T(d) > 0 then
5:             if MC(d) > MC(d*) then
6:                 d* ← d
7:     return d*
```

Although the loop in Algorithm 3 iterates over all k-dimensional vectors over the natural numbers, we will see later that only a polynomial number of vectors need be considered.

Algorithm 3 gives us a distance vector, but not the desired reconciliation at this distance. However, we can augment the algorithm for computing reconciliation counts (given in the next section) to record one such reconciliation. That is, we can extend the reconciliation counts algorithm with annotations such that $Count_{T,v}(\mathbf{d}) = (n, R)$ if and only if there are n distinct reconciliations at distance $\mathbf{d}$ from T which contain event v, and R is one such reconciliation. Therefore, a polynomial time method for computing reconciliation counts allows us to find a 2-approximation to k-centers for MPRs.

4   Computing the Reconciliation Count Function

In this section, we give an algorithm for computing the reconciliation count function, $Count_{T,v} : \mathbb{N}^k \to \mathbb{N}$. The algorithm uses three basic operations: sum, convolution, and shift. Let f and g be functions from $\mathbb{Z}^k$ to $\mathbb{N}$. The *sum*, $h = f + g$, is defined by

$$h(\mathbf{n}) = f(\mathbf{n}) + g(\mathbf{n})$$

The *convolution*, $h = f * g$ is defined by

$$h(\mathbf{n}) = \sum_{\mathbf{m} \in \mathbb{Z}^k} f(\mathbf{m})g(\mathbf{n} - \mathbf{m})$$

The *shift*, $h = f \gg \mathbf{m}$ for some $\mathbf{m} \in \mathbb{Z}^k$ is defined by

$$h(\mathbf{n}) = f(\mathbf{n} - \mathbf{m})$$

In addition, for a given $\mathbf{d} \in \mathbb{Z}^k$ the function $\delta_{\mathbf{d}} : \mathbb{Z}^k \to \mathbb{N}$ is defined by

$$\delta_{\mathbf{d}}(\mathbf{d}) = 1 \text{ and } \delta_{\mathbf{d}}(\mathbf{n}) = 0 \text{ for } \mathbf{n} \neq \mathbf{d}$$

Finally, for a vertex v in the reconciliation graph $\mathcal{G}$ and a MPR T, we define the function $D(v, T)$ as follows:

$$D(v, T) = \begin{cases} -1 & \text{if } v \in \mathbb{E}(T) \\ 1 & \text{if } v \notin \mathbb{E}(T) \text{ and } v \in \mathbb{E}(\mathcal{G}) \\ 0 & \text{if } v \notin \mathbb{E}(\mathcal{G}) \end{cases}$$

For $\mathcal{T} = \{T_1, \ldots, T_k\}$, we define $D(v, \mathcal{T}) = [D(v, T_1), \cdots, D(v, T_k)]$.

Note that while the convolution is defined using an infinite summation, in our usage it will only take a polynomial number of non-zero values and will be shown to be computable in polynomial time.

In Algorithm 4 we give three procedures: The main COUNT function for computing $\text{Count}_{\mathcal{T},v}$ which uses procedures SUBCOUNT and SUPERCOUNT. For clarity, these procedures are described recursively, but the implementation (described in the Supplementary Materials) applies dynamic programming to compute the values bottom-up in polynomial time. The algorithm uses the sum, convolution, shift and $\delta_{\mathbf{d}}$ functions described above.

Algorithm 4. Reconciliation Count

1: **procedure** $\text{COUNT}_{\mathcal{T}}(v)$
2:     **return** $(\text{SUBCOUNT}_{\mathcal{T}}(v) * \text{SUPERCOUNT}_{\mathcal{T}}(v)) \gg [\,|\mathbb{E}(T_1)|, \ldots, |\mathbb{E}(T_k)|\,]$
3:
4: **procedure** $\text{SUBCOUNT}_{\mathcal{T}}(v)$
5:     **if** v is a leaf of G **then**
6:         **return** $\delta_{[-1, -1, \ldots, -1]}$
7:     **if** v is a mapping vertex **then**
8:         **for** each child c_i of v **do**
9:             $f_i \leftarrow \text{SUBCOUNT}_{\mathcal{T}}(c_i)$
10:      **return** $\sum_i f_i$
11:     **if** v is an event vertex **then**
12:         **if** v has one child c **then**
13:            **return** $\text{SUBCOUNT}_{\mathcal{T}}(c) \gg D(v, \mathcal{T})$
14:         **else** v has two children c_1, c_2
15:            **return** $(\text{SUBCOUNT}_{\mathcal{T}}(c_1) * \text{SUBCOUNT}_{\mathcal{T}}(c_2)) \gg D(v, \mathcal{T})$
16:
17: **procedure** $\text{SUPERCOUNT}_{\mathcal{T}}(v)$
18:     **if** v is a root of G **then**
19:         **return** $\delta_{[0, 0, \ldots, 0]}$
20:     **if** v is a mapping vertex **then**
21:         **for** each parent p_i of v **do**
22:             $f_i \leftarrow \text{SUPERCOUNT}_{\mathcal{T}}(p) \gg D(p, \mathcal{T})$
23:             **if** v has a sibling s under p_i **then**
24:                 $f_i \leftarrow f_i * \text{SUBCOUNT}_{\mathcal{T}}(s)$
25:      **return** $\sum_i f_i$
26:     **if** v is an event vertex **then**
27:         $p \leftarrow$ the parent of v
28:         **return** $\text{SUPERCOUNT}_{\mathcal{T}}(p)$

Theorem 2. *Given a reconciliation graph $\mathcal{G}$ for a DTL problem instance and a set $\mathcal{T}$ of k MPRs, the procedure $\text{COUNT}_{\mathcal{T}}(v)$ in Algorithm 4 correctly computes the reconciliation count function.*

Let n denote the larger of the size of the gene tree and the size of the species tree in a DTL instance and let k be a fixed constant representing the size of the representative set $\mathcal{T}$.

Theorem 3. *The worst-case time complexity of Algorithm 4 is $O(n^{k+3} \log n)$.*

Theorem 4. *The worst-case time complexity for performing I iterations of the k-medoids algorithm is $O(In^{k+3} \log n)$.*

Theorem 5. *The worst-case time complexity of the k-centers algorithm is $O(n^{k+3} \log n)$.*

Proofs of these theorems as well as experimental results are available in Supplementary Materials at www.cs.hmc.edu/~hadas/supplement.pdf.

5 Conclusion

In this paper we have studied the problem of clustering the exponentially large space of MPRs in order to find a small number of representative reconciliations and to better understand the structure of MPR space. We have defined a *reconciliation count function*, shown that it can be computed in polynomial time, and demonstrated how this function can be used to implement some well-known clustering algorithms. These results can be extended to other clustering algorithms and to reconciliations whose cost is within some bound of MPR cost.

Finally, there are a number of promising directions for future research. First, while the reconciliation count algorithm runs in polynomial time for any constant k, the running time of $O(n^{k+3} \log n)$ becomes impractical for large values of k. While our initial experimental results suggest that small values of k are likely to be of greatest interest, it may be possible to compute the reconciliation count function more efficiently. Second, the reconciliation count function described here appears to have broad utility in implementing other clustering methods and algorithms and in computing a variety of statistics on clusterings of MPR space. While our experimental results demonstrate the viability of clustering, systematic empirical studies are needed to better understand what clusterings can reveal about the structure of MPR space.

Acknowledgements. This work was funded by the U.S. National Science Foundation under Grant Numbers IIS-1419739 and 1433220. Any opinions, findings, and conclusions or recommendations expressed in this material are those of the author(s) and do not necessarily reflect the views of the National Science Foundation. The authors wish to thank Yi-Chieh Wu and Mukul Bansal for valuable advice and feedback.

References

1. Bansal, M.S., Alm, E.J., Kellis, M.: Efficient algorithms for the reconciliation problem with gene duplication, horizontal transfer and loss. Bioinformatics **28**(12), i283–i291 (2012)

2. Bansal, M.S., Alm, E.J., Kellis, M.: Reconciliation revisited: handling multiple optima when reconciling with duplication, transfer, and loss. In: Deng, M., Jiang, R., Sun, F., Zhang, X. (eds.) RECOMB 2013. LNCS, vol. 7821, pp. 1–13. Springer, Heidelberg (2013). doi:10.1007/978-3-642-37195-0_1

3. Charleston, M.A., Perkins, S.L.: Traversing the tangle: algorithms and applications for cophylogenetic studies. J. Biomed. Inform. **39**(1), 62–71 (2006)

4. Conow, C., Fielder, D., Ovadia, Y., Libeskind-Hadas, R.: Jane: a new tool for cophylogeny reconstruction problem. Algorithms Mol. Biol. **5**(1), 16 (2010)

5. David, L.A., Alm, E.J.: Rapid evolutionary innovation during an Archaean genetic expansion. Nature **469**, 93–96 (2011)

6. Doyon, J.P., Scornavacca, C., Gorbunov, K.Y., Szöllősi, G.J., Ranwez, V., Berry, V.: An efficient algorithm for gene/species trees parsimonious reconciliation with losses, duplications and transfers. Comp. Genomics **6398**, 93–108 (2011)

7. Garey, M.R., Johnson, D.S.: Computers and Intractability: A Guide to the Theory of NP-Completeness. W.H. Freeman & Co., New York (1979)

8. González, T.: Clustering to minimize the maximum intercluster distance. Theor. Comput. Sci. **38**, 293–306 (1985)

9. Ma, W., Smirnov, D., Forman, J., Schweickart, A., Slocum, C., Srinivasan, S., Libeskind-Hadas, R.: DTL-RnB: Algorithms and tools for summarizing the space of DTL reconciliations. IEEE/ACM Trans. Comput. Biol. Bioinform. (2016). doi:10.1109/TCBB.2016.2537319

10. Nguyen, T.H., Ranwez, V., Berry, V., Scornavacca, C.: Support measures to estimate the reliability of evolutionary events predicted by reconciliation methods. PLoS ONE **8**(10), e73667 (2013)

11. Park, H.S., Jun, C.H.: A simple and fast algorithm for k-medoids clustering. Expert Syst. Appl. **36**(2), 3336–3341 (2009)

12. Scornavacca, C., Paprotny, W., Berry, V., Ranwez, V.: Representing a set of reconciliations in a compact way. J. Bioinform. Comput. Biol. **11**(02), 1250025 (2013). pMID: 23600816

13. Than, C., Ruths, D., Innan, H., Nakhleh, L.: Confounding factors in HGT detection: statistical error, coalescent effects, and multiple solutions. J. Comput. Biol. **14**(4), 517–535 (2007)

14. To, T.H., Jacox, E., Ranwez, V., Scornavacca, C.: A fast method for calculating reliable event supports in tree reconciliations via pareto optimality. BMC Bioinform. **16**, 384 (2015)

15. Tofigh, A.: Using trees to capture reticulate evolution: lateral gene transfers and cancer progression. Ph.D. thesis, KTH Royal Institute of Technology (2009)

16. Tofigh, A., Hallett, M.T., Lagergren, J.: Simultaneous identification of duplications and lateral gene transfers. IEEE/ACM Trans. Comput. Biol. Bioinform. **8**(2), 517–535 (2011)

Sequence Analysis and Other Biological Processes

CSA-X: Modularized Constrained Multiple Sequence Alignment

T.M. Rezwanul Islam[⊠] and Ian McQuillan

Department of Computer Science,
University of Saskatchewan, Saskatoon, SK, Canada
rezwanul.islam@usask.ca, mcquillan@cs.usask.ca

Abstract. Imposing constraints that influence multiple sequence alignment (MSA) algorithms can often produce more biologically meaningful alignments. In this paper, a modularized program of constrained multiple sequence alignment (CMSA) called CSA-X is created that accepts constraints in the form of regular expressions. It uses arbitrary underlying MSA programs to generate alignments, and is therefore modular. The accuracy of CSA-X with different underlying MSA algorithms is compared, and also with another CMSA program called RE-MuSiC that similarly uses regular expressions for constraints. A technique is also developed to test the accuracies of CMSA algorithms with regular expression constraints using the BAliBASE 3.0 benchmark database. For verification, ProbCons and T-Coffee are used as the underlying MSA programs in CSA-X, and the accuracy of the alignments are measured in terms of Q score and TC score. Based on the results presented herein, CSA-X significantly outperforms RE-MuSiC. On average, CSA-X used with constraints that were algorithmically created from the least conserved regions of the correct alignments achieves results that are 17.65% higher for Q score, and 23.7% higher for TC score compared to RE-MuSiC. Further, CSA-X with ProbCons (CSA-PC) achieves a higher score in over 97.9% of the cases for Q score, and over 96.4% of the cases for TC score. It also shows that the use of regular expression constraints, if chosen well, created from accurate knowledge regarding a lesser conserved region can improve alignment accuracy. Statistical significance is measured using the Wilcoxon rank-sum test and Wilcoxon signed-rank test. An open source implementation of CSA-X is also provided.

Keyword: Multiple sequence alignment

1 Introduction

Multiple sequence alignment (MSA) is a fundamental tool towards many objectives, such as phylogenetic studies, computational biology, prediction of functional residues, and protein structure prediction [17]. A large number of MSA programs have been developed, and Pais, FSM et al. [16] recently surveyed such programs in terms of accuracy and computational time. Indeed, accuracy is

© Springer International Publishing AG 2017
D. Figueiredo et al. (Eds.): AlCoB 2017, LNBI 10252, pp. 143–154, 2017.
DOI: 10.1007/978-3-319-58163-7_10

particularly important for MSA algorithms, especially within modern computer bioinformatics pipelines, where less accurate alignments cause negative downstream effects with amplification of errors [12]. Most of the state-of-the-art multiple sequence alignment programs such as ProbCons [7], T-Coffee [15], MAFFT [11], and ClustalW [20] are fully automated, with a limited number of changeable parameters.

But often, users have additional information that could affect the alignments such as, active site residues, intramolecular disulphide bonds, enzyme activities, and conserved motifs [19]. Hence, having a program that can use additional information, either manually created, or automatically determined from additional annotations, can improve the accuracy of alignments. Constrained multiple sequence alignment (CMSA) [18] is an extension of the MSA problem [4] that allows users to use knowledge regarding the sequences involved, in the form of constraints, with a view to achieving more biologically meaningful alignments. For example, Du and Lin [8] showed that ClustalW [20], does not align common patterns and similar structures found in sequences consistently. Because of this reason, Tsai et al. [19] proposed MuSiC, a web server that allowed constrained alignment of sequences. But many biologically important motifs, such as those listed as regular expressions in the PROSITE [10] database cannot be formulated into constraints according to the convention followed by MuSiC [5]. To solve this issue, Arslan [2], and Chung et al. [6] introduced alignment algorithms that accept regular expression constraints, and enforce that segments that match the regular expression must align. Then, Chung et al. proposed RE-MuSiC [5], an extension to their previous work [6] to support multiple sequences and multiple constraints. In that work, they used sequence motifs found in PROSITE as regular expression constraints to improve the quality of alignments. However, there are some limitations of RE-MuSiC, as it does not allow the use of quantification operators such as Kleene star (*), Kleene plus (+) in regular expression constraints, and thus only a subset of regular expressions can be used as input. Arslan [3] also proposed sequence alignment programs guided by Context Free Grammars (CFG) only limited to pairwise sequence alignment. Morgenstern et al. [14] developed DIALIGN, a web server, that can accept user defined anchor points as constraints. It is common to use a benchmark database, such as BALiBASE to evaluate MSA algorithms [9,11,15], however no such technique is available for CMSA with regular expression constraints, and therefore, no such comparison is available for use with RE-MuSiC.

Here a new program, CSA-X is developed that also accepts arbitrary regular expression constraints (including quantifiers), and creates a multiple sequence alignment that forces sections to align that match the entire regular expression. Furthermore, it is also possible to enforce with an extended regular expression syntax that certain sections that match part of a regular expression must align. CSA-X is a modularized program that uses an underlying MSA program, and because of this reason, it is possible to replace the underlying MSA program with another, perhaps improved or tailored program. In addition, this study compares the performance of CSA-X, RE-MuSiC, and 'X', where 'X' is the underlying MSA program in the proposed tool, with respect to the BALiBASE 3.0 [21] benchmark database. This involves the creation of a new technique to compare CMSA algorithms by creating regular expressions algorithmically using the BALiBASE alignments. This assesses their accuracy in terms of Q score and TC score [9], and measures statistical significance of the results. Furthermore, it also shows that if constraints are chosen appropriately, such as from knowledge regarding lesser conserved regions, CSA-X can give better results than the underlying MSA algorithm.

2 Methods

An open source implementation of CSA-X has been made available [1], which will be described. Arbitrary MSA implementation can be used with it. CSA-X accepts constraints in the form of regular expressions using the PERL regular expression syntax. However, the symbol # can be optionally placed in multiple spots in the regular expression, and it has special meaning providing guidance by which sequences are aligned. Next the constraints are defined.

Definition 1. *Hash-augmented regular expressions are defined inductively:*

- *every PERL regular expression is a hash-augmented regular expression,*
- *if R and S are two hash-augmented regular expressions, then R#S is a hash-augmented regular expression.*

From this definition, it is implied that every hash-augmented regular expression can be written in the following form, for some $n \geq 1$: $R_1\#R_2\#\ldots\#R_n$, when $R_1,\ldots,R_n$ are regular expressions. Intuitively, the MSA generated will align the parts of each sequence that match $R_1, R_2, \ldots, R_n$, and enforce that the parts that match each R_i, for $1 \leq i \leq n$ are aligned. Hence, the # symbols provide additional control by giving information regarding the residues or nucleotides to align. If the # symbols are not used, then CSA-X constructs the best alignment of the entire parts matching the entire regular expression.

In the case of hash-augmented regular expressions, between every two # symbols, it must be a syntactically correct regular expression. For example, (AC#TT)C#A is not a valid CSA-X hash-augmented regular expression because the left side of the first # symbol contains '(AC' and right side contains 'TT)C' which are not regular expressions.

If a hash-augmented regular expression matches multiple sequences, then each match must have the same number of hash symbols since # symbols cannot (by definition) be placed inside any quantifier, such as * (which could match i times within one sequence, but j times within another, where $i \neq j$).

Consider, an input of N sequences to align $S_1, S_2, S_3, \ldots, S_N$, $N \geq 2$, and a hash-augmented regular expression $R = R_1\#R_2\#R_3\#\ldots\#R_m$, where $m \geq 1$. The precise process that CSA-X uses to generate such an alignment can be described using the following high-level steps:

1. CSA-X attempts to match R to each sequence S_i of the N input sequences, for each i, $1 \leq i \leq N$. If the regular expression matches exactly once in each sequence S_i, then CSA-X determines a list of positions (for each i, $1 \leq i \leq N$) $l_i^0, l_i^1, l_i^2, \ldots, l_i^m$, where $0 \leq l_i^0 \leq l_i^1 \leq l_i^2 \leq \ldots \leq l_i^m \leq |S_i|+1$, whereby regular expression R_j matches between positions l_i^{j-1} and $l_i^j - 1$, for each j, $1 \leq j \leq m$ (if $l_i^{j-1} = l_i^j$ then R_j matches the empty string). If CSA-X does not find any regular expression matches on the input sequences or it finds matches for a strict subset of sequences in the dataset, then it returns the alignment of the input dataset using the underlying MSA program without using the regular expression.

2. CSA-X generates alignments for each of the matched sections of the sequences using the underlying alignment algorithm. That is, it aligns the sub-words $S_1(1, l_1^0 - 1), \ldots, S_N(1, l_N^0 - 1)$, then aligns subwords $S_1(l_1^{j-1}, l_1^j - 1), \ldots, S_N(l_N^{j-1}, l_N^j - 1)$ for every j, $1 \leq j \leq m$, and then aligns subwords $S_1(l_1^{m+1}, |S_1|), \ldots, S_N(l_N^{m+1}, |S_N|)$. Then it concatenates each of these alignments together in order. As the constraints in CSA-X are specified using a hash-augmented regular expression, it generates alignments by decoding information from the specified constraints.

3. If CSA-X finds multiple regular expression matches on a single sequence, then it generates all possible combinations of the matched-segment datasets by selecting each regular expression match of the sequence separately, and determines the alignment that has the highest sum-of-pairs score.

Intuitively, the formalism of step 2 means that CSA-X identifies the segments that match the entire expression R for each sequence S_i in step 1. In addition, at the same time on each of the matched segments, it also identifies the subsections that match the sub-patterns $R_1, R_2, R_3, \ldots, R_m$ consecutively in the hash-augmented regular expression. Then, CSA-X aligns each matching sub-pattern separately (including the parts that match before the first matching sub-pattern, and the parts that match after the final sub-pattern has ended), using the underlying MSA program X to generate alignments, and then it merges the generated alignments together to produce a complete alignment.

In step 3, suppose a hash-augmented regular expression R, matches the sequence S_t at two spots. If the rest of the sequences match the hash-augmented regular expression exactly at one spot, then CSA-X would create two alignments, one where the matching occurs between the first matching part of S_t, and the other one where the matching occurs with the second matching part of S_t. Then the algorithm determines the alignment that has the highest sum-of-pairs score, and returns the alignment with the highest score.

It should be noted that, multiple regular expressions R and S can be used as input by joining them with quantifiers such as $R.^*S$, where "." represents any character match, and "*" is Kleene star. Alignment can be further influenced through # symbols, e.g. $R\#.^*\#S$.

Example 2. Conserved motifs for different protein sequences are listed in the PROSITE database in the form of regular expressions, which can be used as constraints to improve the biological accuracy of the alignments in different CMSA programs. For example, the TATA-binding protein plays a vital role in the activation of eukaryotic genes. PROSITE (PDOC00303) lists the consensus for the signature pattern of the TATA-binding protein as follows (using a slightly different regular expression syntax).

Y-x-[PK]-x(2)-[IF]-x(2)-[LIVM](2)-x-[KRH]-x(3)-P-[RKQ]-x(3)-L-
[LIVM]-F-x-[STN]-G-[KR]-[LIVMA]-x(3)-G-[TAGL]-[KR]-x(7)-
[AGCS]-x(7)-[LIVMF].

For the alignment of different TATA box proteins, the above mentioned consensus pattern can be used as a constraint. For instance, if one would like to align TATA box proteins found in *Homo sapiens* (gb AAI09054.1), *Rattus rattus* (gb AAH16476.1), and microorganism *Halobacterium salinarum* (emb CAA63691.1) using CSA-X, then the format of this consensus pattern can be as follows:

Y.[PK]..[IF]..[LIVM]{2}.[KRH]...P[RKQ]...L[LIVM]F.[STN]G
[KR][LIVMA]...G[TAGL][KR].......[AGCS].......[LIVMF].

It is also possible to simplify this further using quantifiers, or to add hash symbols to force sections of the regular expression to align; for instance to align sections of the sequences that match `Y.[PK]..[IF]..[LIVM]+.[KRH]`, then `...P[RKQ]...L [LIVM]F.[STN]G[KR][LIVMA]`, then `...G[TAGL][KR].......[AGCS]....... [LIVMF]` the hash augmented regular expression could be used. Omitting the # symbols would not necessarily align the three parts separately on all sequences.

Y.[PK]..[IF]..[LIVM]+.[KRH]#...P[RKQ]...L[LIVM]F.[STN]G[KR]
[LIVMA]#...G[TAGL][KR].......[AGCS].......[LIVMF].

Figure 1 shows the partial alignment generated by CSA-X, where the region identified by the regular expression is aligned in columns (highlighted). For this alignment, ProbCons is used as the underlying alignment tool.

3   Method of Assessments

Since CSA-X is a modular tool, the underlying MSA program can be changed to obtain different alignments, and in some sense, different customized tools. The study conducted by Pais, FSM et al. [16] showed that ProbCons, T-Coffee, Probalign, and MAFFT achieve higher accuracy than other MSA tools considered. Therefore, for this assessment, ProbCons and T-Coffee are used as the underlying MSA algorithms in CSA-X (although other programs can be used with CSA-X as well, these are the only two used for the purposes of assessment). Whenever CSA-X uses ProbCons, it is referred to as CSA-PC, and for T-Coffee, it is called CSA-TCOF. It is common for a benchmark database, such as BALiBASE 3.0 to be used to assess alignments and algorithms. Each set of sequences in such a database also has an alignment that is thought of as "correct" (based on additional knowledge such as protein structure). This database is

```
            201                                              250
  gi80478871 .......................YNPKRFAAVIMRIREPRTTALIFSS
  gi16741283 MRIREPRTTALIFSSGKMVCTGAKSYEPELFPGLIYRMIKPRIVLLIFVS
  gi1070345  .......................YNPEDFPGVVYRLQEPKSATLIFRS

            251                                              300
  gi80478871 GKMVCTGAKSEEQSRLAARKYARVVQKLGFPAK.FLDFKIQNMVGSCDVK
  gi16741283 GKVVLTGAKVRAEIYEAFENIYPIL.........................
  gi1070345  GKVVCTGAKSVDDVHEALGIVFGDIRELGIDVTSNPPIEVQNIVSSASLE

            301                                              350
  gi80478871 FPIRLEGLVLTH.QQFSSYEPELFPGLIYRMIKPRIVLLIFVSGKVVLTG
  gi16741283 ..................................................
  gi1070345  QSLNLNAIAIGLGLEQIEYEPEQFPGLVYRLDDPDVVVLLFGSGKLVITG
```

Fig. 1. CSA-X partial alignment of TATA box proteins. CSA-X partial alignment of TATA box proteins, where the highlighted regions indicate the sections matched by the regular expression constraint.

used here. However, RE-MuSiC generates erroneous alignments for a portion of the datasets in the BAliBASE 3.0 benchmark database, where the length of the sequences are not equal (sometimes the resulting alignments produced contain wildcard characters). Hence, the *working database* for this study is defined as being created from BAliBASE 3.0 including those datasets for which RE-MuSiC produces non-erroneous alignments for the purposes of comparison. BAliBASE 3.0 is classified into several groups; namely RV11, RV12, RV20, RV30, RV40, and RV50. In the working database for this study, there are 76 datasets from RV11, 84 datasets from RV12, 6 datasets from RV20, 6 datasets from RV30, 17 datasets from RV40 and 11 datasets from RV50; in total 200 datasets from BAliBASE 3.0 out of a total of 386 datasets. Out of these 200 datasets, 98 datasets contain short truncated sequences. To compare the performance of CSA-X with RE-MuSiC and other programs, this working database is used in this study (we will additionally consider the difference in results between programs when including those datasets not in the working dataset).

As BAliBASE does not contain regular expression constraints, a new technique must be developed for comparison of CMSA algorithms. To identify the effects of constraints on generated alignments, two sets of regular expression constraints are created to use for assessment. Indeed, the correct alignments from the BAliBASE 3.0 benchmark database are used to algorithmically create accurate regular expressions. One set of regular expression constraints are created from the "most conserved" region of the correct alignments. Another set is constructed from the "least conserved" region of the correct alignments. All of these constraints are automatically generated using a Perl script, which uses reference alignment files from BAliBASE 3.0, and identifies the most conserved regions and the least conserved regions for the alignments and generates the regular expression constraints. These same sets of constraints are used to compare CSA-X and RE-MuSiC.

The idea behind this approach is to identify the effects of constraints on multiple sequence alignment. Often, expert users possess information about the sequences involved in the alignment process. They align the sequences using a MSA program, and then often adjust the alignment based on knowledge not reflected in the generated alignment. For the assessment, we do not obtain regular expression constraints from expert users, but rather, algorithmically create regular expressions from the curated alignments. Although this approach could create unrealistically good regular expressions, it is useful to use when comparing multiple algorithms that take regular expressions as constraints and to represent "accurate" knowledge being incorporated into regular expressions.

For this study, the regular expression constraints are generated to be of length 12 with a maximum of one gap per sequence, which is large enough to affect alignments, while avoiding many matches. To make a fair comparison between CSA-X and RE-MuSiC, both are tested on the same sets separately for both (most and least conserved) of regular expression constraints, and therefore, all regular expressions tested do not have the quantifiers * or + as these do not work with RE-MuSiC. Furthermore, a separate comparison is made between CSA-PC with these regular expressions and ProbCons without using any regular expressions at all (and similarly with T-Coffee) to gauge the potential improvements that using regular expressions as constraints can provide. This depends on whether the regular expressions are created from highly conserved or lesser conserved regions. Although this part of the assessment is done using the correct alignments to construct the regular expressions, it is only being used to see if regular expressions can possibly improve quality, depending on the type of regular expression. A thorough test of common regular expressions used by expert users together with a test to see if they improve alignment quality would be valuable. However, for comparing RE-MuSiC to CSA-X, such regular expressions are equally favourable to both programs, and is therefore a useful method of comparison.

3.1   Accuracy, Statistical Significance and Parameters

To measure the accuracy of considered programs in this study, two scores, Q score (Quality Score) and TC score (Total Column Score) are computed. Edgar [9] defined the Q score of an algorithm as a ratio between the number of correctly aligned pairs to the number of residue pairs in the reference alignment. This is the same as the sum-of-pairs score defined by Thompson et al. [22]. TC score is the number of correctly aligned columns, divided by the number of columns in the reference alignment (this is the same as the column score (CS) defined by Thompson et al.).

To reduce the probability that the difference is merely by chance, researchers working in the area of MSA frequently conduct statistical significance tests. In this work, Wilcoxon signed-rank test [23] and Wilcoxon rank-sum test [23] are used to measure statistical significance. If two samples are paired, Wilcoxon signed-rank test is used, otherwise, Wilcoxon rank-sum test is used.

Standalone ProbCons and T-Coffee are used with the default parameter settings (performing a comparison by systematically varying all parameters with every program is of interest but is left for future work). The same parameter settings of ProbCons and T-Coffee are used in CSA-PC and CSA-TCOF respectively. RE-MuSiC is run with the default gap extension and gap open penalty. CSA-PC, CSA-TCOF, and RE-MuSiC are provided with the equivalent set of regular expression constraints. As the format of specifying regular expression constraints in CSA-X and RE-MuSiC is different, equivalent regular expression constraint sets are used for these programs.

4   Results

For each of CSA-PC, CSA-TCOF, RE-MuSiC, T-Coffee, and ProbCons, average (AVG) and standard deviation (SD) of Q score and TC score are presented in Table 1. Among these programs CSA-PC, CSA-TCOF, and RE-MuSiC are provided with the regular expression constraints; however, T-Coffee and ProbCons are used without any constraints (as they do not take any as input).

4.1   Comparison of CSA-X with RE-MuSiC

It is observed from Table 1 that for the 200 datasets in the working database, that CSA-PC and CSA-TCOF both achieve higher accuracy compared to RE-MuSiC, using both Q score and TC score. From Table 1 it can be calculated that

Table 1. Average and standard deviation of Q score and TC score for the working database. MC (and LC respectively) represent the use of regular expression constraints identified from the most conserved region (least conserved respectively), of the correct alignments in the benchmark datasets ('−' represents a score that cannot be computed). The entries that are bold represent the highest value for each type of score and regular expression.

		MC	LC	AVG (SD)
		AVG (SD)	AVG (SD)	
Q	CSA-PC	**0.868 (0.118)**	**0.881 (0.116)**	−
	CSA-TCOF	0.860 (0.131)	0.876 (0.124)	−
	RE-MuSiC	0.691 (0.197)	0.702 (0.220)	−
	ProbCons	−	−	0.854 (0.153)
	T-Coffee	−	−	0.846 (0.166)
TC	CSA-PC	**0.713(0.222)**	**0.730 (0.244)**	−
	CSA-TCOF	0.702 (0.231)	0.718 (0.244)	−
	RE-MuSiC	0.496 (0.256)	0.487 (0.299)	−
	ProbCons	−	−	0.693
	T-Coffee	−	−	0.680

on average for Q score, CSA-PC achieves approximately 0.179 and CSA-TCOF achieves almost 0.174 higher score compared to RE-MuSiC when using constraints obtained from the least conserved (LC) region of the correct alignments respectively. For the constraints obtained from the most conserved (MC) region of the correct alignments, CSA-PC and CSA-TCOF achieve 0.176 and 0.168 higher score respectively. While for TC score, for the most conserved region regular expression constraints set, CSA-PC achieves 0.217 and CSA-TCOF achieves 0.206 higher results compared to RE-MuSiC, and the score rises by 0.217 and 0.206 respectively for CSA-PC and CSA-TCOF for the LC constraints set.

Out of 200 working datasets for Q score, CSA-PC and CSA-TCOF with LC constraints perform higher for 195 and 194 datasets respectively compared to RE-MuSiC. In addition, for TC score, CSA-PC and CSA-TCOF with LC constraints set achieves higher score in total for 185 and 184 datasets. CSA-PC and CSA-TCOF with MC constraints set achieves a higher score for Q score for 192 and 191 datasets respectively, and for TC score they achieve higher score for 186 and 180 datasets respectively compared to RE-MuSiC. In addition, if all the datasets in BAliBASE 3.0 are considered, instead of just the working datasets, and we define CSA-X as performing better for instances where RE-MuSiC is giving erroneous results, then CSA-PC (LC) gives a higher score for 381 datasets out of 386 datasets, and CSA-TCOF (LC) gives a higher score for 380 datasets out of 386 datasets.

4.2 Comparison of CSA-X with the Underlying MSA Program

From Table 1, CSA-PC and CSA-TCOF score higher overall than standalone ProbCons and T-Coffee respectively run without any constraints. For MC constraints, CSA-PC and CSA-TCOF both show 0.014 higher Q score and more than 0.02 higher TC score compared to ProbCons and T-Coffee. Further, using LC constraints, CSA-PC and CSA-TCOF achieve 0.026 and 0.029 higher Q score and 0.0378 and 0.0376 higher TC score respectively. According to Thompson et al. [22] the BAliBASE sum-of-pairs score (similar to Q score) increases if a program succeeds in aligning sequences relative to the reference alignment dataset; this means that the higher the Q score is, the better the program is at generating accurate alignments, while TC score tests how efficiently the program is aligning all the sequences. This is a more stringent criteria of measurement as a column score can become zero if a single sequence is misaligned [13]. Again, it is worth mentioning that the comparison of CSA-X to its underlying tool does not lend any evidence to the notion that CSA-X is better than its underlying algorithm. It is only examining certain types of correct information that can improve alignments. Furthermore, it is an important verification of the potential of CSA-X.

4.3 Statistical Analysis

First, the Wilcoxon rank-sum test is performed between CSA-TCOF and RE-MuSiC, and between CSA-PC and RE-MuSiC. As the outcome of CSA-TCOF does not depend upon RE-MuSiC, the Wilcoxon rank-sum test is chosen. Second, each are compared to their underlying algorithm with both the most conserved and least conserved regular expressions. Since the outcome of CSA-TCOF and CSA-PC depends upon T-Coffee and ProbCons respectively, the Wilcoxon signed-rank test is selected to test if there is significant difference between these programs. Table 2 shows the results of these tests. For both the tests, the null hypothesis is that there is no significant difference between the two samples. If the test rejects the null hypothesis then it means that there is a significant difference between the two samples. For this test, a 5% significance level is used, which means that if the p-value is less than 0.05 then the null hypothesis is rejected. For the Wilcoxon rank-sum test, all the p-values are significantly less than 0.05. Hence, the null hypothesis is rejected, and it is determined that the results of CSA-PC and CSA-TCOF are significantly different compared to RE-MuSiC, and the differences are not by chance. However, for the Wilcoxon signed-rank test, the results are not significantly different for CSA-TCOF and T-Coffee if CSA-TCOF uses the most conserved (MC) regular expression constraints set, as with CSA-PC. This is because ProbCons and T-Coffee both are able to successfully align the most conserved region without the explicit constraints. But the situation changes if CSA-TCOF and CSA-PC uses the least conserved (LC)

Table 2. Wilcoxon rank-sum and Wilcoxon signed-rank test results.

Wilcoxon rank-sum test			
Constraints	Programs	Scores	P-value (Significant)
MC	CSA-TCOF and RE-MuSiC	Q	<2.2e-16 (yes)
		TC	2.89e-15 (yes)
	CSA-PC and RE-MuSiC	Q	<2.2e-16 (yes)
		TC	<2.2e-16 (yes)
LC	CSA-TCOF and RE-MuSiC	Q	<2.2e-16 (yes)
		TC	6.66e-16 (yes)
	CSA-PC and RE-MuSiC	Q	<2.2e-16 (yes)
		TC	<2.2e-16 (yes)
Wilcoxon signed-rank test			
Constraints	Programs	Scores	P-value (Significant)
MC	CSA-TCOF and T-Coffee	Q	<0.6529 (no)
		TC	0.1579 (no)
	CSA-PC and ProbCons	Q	<0.0911 (no)
		TC	0.0201 (yes)
LC	CSA-TCOF and T-Coffee	Q	<8.18e-10 (yes)
		TC	3.09e-08 (yes)
	CSA-PC and ProbCons	Q	<2.50e-10 (yes)
		TC	<1.10e-08 (yes)

regular expression constraints set, and it is observed that there is significant difference in the results of CSA-TCOF and CSA-PC with T-Coffee and ProbCons respectively if they are supplied with LC constraints set.

Although, constraints chosen from the most or least conserved region are not necessarily realistic in terms of regular expression constraints chosen by either an expert user, or created from a database of additional information, using constraints created from the correct alignment does have the advantage of capturing some piece of information the user may know to be true, in a situation where a standalone alignment program is not giving the desired results. And indeed, constraints chosen from the most conserved region do not seem to help significantly versus not using any constraint, however constraints chosen from the least conserved region do help versus not using any constraints.

5   Conclusions

The constrained multiple sequence alignment program, CSA-X, allows the user to specify regular expression constraints for the multiple sequence alignment, and the resulting alignment enforces that specific sections matching the regular expression gets aligned. This can improve the accuracy and biological significance of the generated alignments, as functional and structural information regarding the sequences can be expressed using regular expression syntax. However, a more systematic study of regular expression constraints from expert users and other sources is left as future work.

In this research work, based on the average accuracy scores from the benchmarking analysis and the statistical significance testing, it is shown that CSA-X framework with ProbCons and T-Coffee (known as CSA-PC and CSA-TCOF respectively) generates more accurate alignments compared to RE-MuSiC—the only other implemented CMSA algorithm that uses regular expression constraints. Furthermore, it is also shown that if good regular expression constraints are chosen from the least conserved portion of the correct alignments, then the results of CSA-X are significantly better than the underlying MSA program. Finally, CSA-X is a modularized tool, and it allows the user to change the underlying multiple sequence alignment program if more efficient programs become available, or a specialized program is required. An open source implementation is also available [1].

References

1. CSA-X. https://bitbucket.org/RezwanIslam/csa-x/downloads. Accessed 28 Jan 2017
2. Arslan, A.N.: Regular expression constrained sequence alignment. J. Discrete Algorithms **5**(4), 647–661 (2007)
3. Arslan, A.N.: Sequence alignment guided by common motifs described by context free grammars. In: Biotechnology and Bioinformatics Symposium (BIOT) (2007)
4. Carrillo, H., Lipman, D.: The multiple sequence alignment problem in biology. SIAM J. Appl. Math. **48**(5), 1073–1082 (1988)

5. Chung, Y.S., Lee, W.H., Tang, C.Y., Lu, C.L.: RE-MuSiC: A tool for multiple sequence alignment with regular expression constraints. Nucleic Acids Res. **35**(suppl 2), W639–W644 (2007)
6. Chung, Y.-S., Lu, C.L., Tang, C.Y.: Efficient algorithms for regular expression constrained sequence alignment. In: Lewenstein, M., Valiente, G. (eds.) CPM 2006. LNCS, vol. 4009, pp. 389–400. Springer, Heidelberg (2006). doi:10.1007/11780441_35
7. Do, C.B., Mahabhashyam, M.S., Brudno, M., Batzoglou, S.: ProbCons: probabilistic consistency-based multiple sequence alignment. Genome Res. **15**(2), 330–340 (2005)
8. Du, Z., Lin, F.: Pattern-constrained multiple polypeptide sequence alignment. Comput. Biol. Chem. **29**(4), 303–307 (2005)
9. Edgar, R.C.: MUSCLE: multiple sequence alignment with high accuracy and high throughput. Nucleic Acids Res. **32**(5), 1792–1797 (2004)
10. Hulo, N., Bairoch, A., Bulliard, V., Cerutti, L., De Castro, E., Langendijk-Genevaux, P.S., Pagni, M., Sigrist, C.J.: The PROSITE database. Nucleic Acids Res. **34**(suppl 1), D227–D230 (2006)
11. Katoh, K., Misawa, K., Kuma, K., Miyata, T.: MAFFT: a novel method for rapid multiple sequence alignment based on fast fourier transform. Nucleic Acids Res. **30**(14), 3059–3066 (2002)
12. Kumar, S., Filipski, A.: Multiple sequence alignment: in pursuit of homologous DNA positions. Genome Res. **17**(2), 127–135 (2007)
13. Lassmann, T., Sonnhammer, E.L.: Kalign — an accurate and fast multiple sequence alignment algorithm. BMC Bioinformatics **6**(1), 298 (2005)
14. Morgenstern, B., Werner, N., Prohaska, S.J., Steinkamp, R., Schneider, I., Subramanian, A.R., Stadler, P.F., Weyer-Menkhoff, J.: Multiple sequence alignment with user-defined constraints at GOBICS. Bioinformatics **21**(7), 1271–1273 (2005)
15. Notredame, C., Higgins, D.G., Heringa, J.: T-Coffee: a novel method for fast and accurate multiple sequence alignment. J. Mol. Biol. **302**(1), 205–217 (2000)
16. Pais, F.S.-M., de Ruy, P., Oliveira, G., Coimbra, R.S.: Assessing the efficiency of multiple sequence alignment programs. Algorithms Mol. Biol. **9**(1), 1–4 (2014). BioMed Central
17. Papadopoulos, J.S., Agarwala, R.: COBALT: constraint-based alignment tool for multiple protein sequences. Bioinformatics **23**(9), 1073–1079 (2007)
18. Tang, C.Y., Lu, C.L., Chang, M.D.T., Tsai, Y.T., Sun, Y.J., Chao, K.M., Chang, J.M., Chiou, Y.H., Wu, C.M., Chang, H.T., Chou, W.I.: Constrained multiple sequence alignment tool development and its application to RNase family alignment. J. Bioinform. Comput. Biol. **1**(02), 267–287 (2003)
19. Te Tsai, Y., Huang, Y.P., Yu, C.T., Lu, C.L.: MuSiC: a tool for multiple sequence alignment with constraints. Bioinformatics **20**(14), 2309–2311 (2004)
20. Thompson, J.D., Higgins, D.G., Gibson, T.J.: CLUSTAL W: improving the sensitivity of progressive multiple sequence alignment through sequence weighting, position-specific gap penalties and weight matrix choice. Nucleic Acids Res. **22**(22), 4673–4680 (1994)
21. Thompson, J.D., Koehl, P., Ripp, R., Poch, O.: BAliBASE 3.0: latest developments of the multiple sequence alignment benchmark. Proteins: Struct., Funct., Bioinf. **61**(1), 127–136 (2005)
22. Thompson, J.D., Plewniak, F., Poch, O.: A comprehensive comparison of multiple sequence alignment programs. Nucleic Acids Res. **27**(13), 2682–2690 (1999)
23. Triola, M.M., Triola, M.F.: Biostatistics for the Biological and Health Sciences. Pearson Addison-Wesley, Boston (2006)

Quantifying Information Flow
in Chemical Reaction Networks

Ozan Kahramanoğulları[1,2]

[1] Department of Mathematics, University of Trento, Trento, Italy
[2] The Micrososft Research - University of Trento
Centre for Computational and Systems Biology, Rovereto, Italy

Abstract. We introduce an efficient algorithm for stochastic flux analysis of chemical reaction networks (CRN) that improves our previously published method for this task. The flux analysis algorithm extends Gillespie's direct method, commonly used for stochastically simulating CRNs with respect to mass action kinetics. The extension to the direct method involves only book-keeping constructs, and does not require any labeling of network species. We provide implementations, and illustrate on examples that our algorithm for stochastic flux analysis provides a means for quantifying information flow in CRNs. We conclude our discussion with a case study of the biochemical mechanism of gemcitabine, a prodrug widely used for treating various carcinomas.

Keywords: Chemical reaction networks · Stochastic simulation · Flux

1 Introduction

Chemical reaction networks (CRNs) provide a convenient representation scheme for a broad variety of models in biology and ecology. By resorting to mass action kinetics, CRNs can be simulated deterministically or stochastically. Stochastic simulations are commonly performed by using Gillespie's direct method [5], or its extensions that address a variety of concerns such as efficiency, e.g., [7], computation of rare events, e.g., [11], or portability, e.g., [2].

While it is now common practice to use deterministic and stochastic simulations interchangeably for a given CRN as well as hybrid simulations [16], these methods provide their own merits in different settings. Deterministic simulations root in a rich theory that makes available various analysis techniques, including flux analysis [15], as well as efficient numerical methods that also ease practical tasks such as model fitting by linear regression. However, differential equation simulations provide only approximations of the changes in population sizes of CRNs, as random fluctuations cannot be retrieved without introducing an additional machinery on top of deterministic methods. In this respect, because it is practically implausible to directly obtain the solution of the chemical master equation (CME) [6,13] for a not-extremely-small CRN, stochastic simulation

© Springer International Publishing AG 2017
D. Figueiredo et al. (Eds.): AlCoB 2017, LNBI 10252, pp. 155–166, 2017.
DOI: 10.1007/978-3-319-58163-7_11

algorithms come in handy for computing stochastic trajectories of CRNs. Nevertheless, there are recent efforts that address ways for pushing the envelope by using linear approximations of CME for stochastic analysis of CRNs [1].

In previous work [10], we have presented a method for stochastic flux analysis of CRNs that is based on a consideration of stochastic simulations with CRNs as non-interleaving computations of concurrent systems [8,14]. In this approach, a simulation is considered as a reduction of a complex structure, that is, the CRN, into a simpler structure, that is, the simulation trajectory. When a simulation trajectory is read as a time series, the reaction instances are totally ordered, because the time stamps of the reaction instances specify a sequential order on them. However, when the reaction instances are considered from the point of view of their causal dependencies with respect to their production and consumption relationships with each other, the simulation trajectory takes a partial order structure rather than a sequential total order structure. In order to retrieve this otherwise lost information, the algorithm in [10] labels each species instance with a unique identifier, thereby making it possible to trace each species instance during the simulation. By tracing these identifiers, the method constructs a partial order structure of species instances. This structure is then used to quantify the causal interdependence of the reaction instances, and compressed to reveal the flux graph after a number of graph transformations.

The method described above introduces a departure from the Gillespie's direct method, as this algorithm is not designed to trace individual species instances, but rather monitor the network state as a vector of species types. Although monitoring each species does not increase the complexity of the Gillespie algorithm or hamper its correctness, it introduces an overhead due to the individual representation of species. This overhead effects the simulation efficiency, and extends the simulation time in comparison to the standard Gillespie algorithm. In some cases with tens of thousands of individuals, it also introduces a limiting factor for running simulations due to the memory required to trace individuals.

In the following, we introduce an efficient algorithm for stochastic flux analysis of CRNs in the form of a simple extension of Gillespie's direct method. In this algorithm, the fluxes of a CRN are computed during simulation by updating two arrays, the size of which are bounded by the number of reactions and species-types of the CRN. Such a mechanism of book-keeping makes it possible to monitor the network state during simulation in the form of a species-type vector as in the direct method. Consequently, the algorithm computes the flux graphs without being subject to an overhead due to monitoring of the species.

As in [10], the flux graph can be extracted for any time interval, in steady or stationary state, and it provides a causality summary of the network resources, resulting in a quantification of the information flow in the simulation. We illustrate our method on experiments with example networks. We conclude our discussion with a case study of the biochemical mechanism of gemcitabine, a prodrug widely used for treating various carcinomas. The flux graphs of this network visualize how system dynamics is affected in different metabolic regimes.

The modules, including a tool for computing flux paths, and the examples below are available for download at our website.[1]

2   Stochastic Simulation and Flux Analysis

The stochastic flux analysis of chemical reaction networks [10] is a general method on discrete event systems that can be represented as Markov chains. Such systems include those that implement mass action kinetics, which are used in systems biology for modeling a broad spectrum of phenomenon from those in molecular biology to large ecosystems. We thus here focus on chemical reaction networks (CRN) as they are studied in systems biology. Specifically, we use those that are commonly simulated by the Gillespie algorithm [5], and can be approximated by deterministic ordinary differential equation systems. The methods we discuss below, however, can be generalized to systems that are represented as discrete event systems. We first review CRNs, and stochastic flux analysis. We refer to [10] for the technical definitions and examples that are not included here.

A CRN consists of a set of reactions and an initial state. A reaction

$$m_1 R_1 + \ldots + m_l R_l \xrightarrow{\rho} n_1 P_1 + \ldots + n_r P_r$$

describes the species $R_1, \ldots, R_l$ that reaction consumes when it occurs, and the species $P_1, \ldots, P_r$ that it produces. The constants $m_1, \ldots, m_l$ and $n_1, \ldots, n_r$ are positive integers that denote the multiplicity of the reactants that are consumed and the products that are produced, respectively, at every instance of such a reaction. The reaction rate constant ρ is a positive real number, which determines how often a reaction occurs in a system, depending on the availability of the reactants that the reaction consumes. According to the mass action kinetics, the probability of a reaction's firing at a particular state instead of another is proportional with the multiplication of ρ and the number of possible combinations of reactants at that state. In this respect, the initial state can be safely considered as a special reaction with infinite rate, which consumes a dummy species, e.g., *Init*, which is always present at the beginning of the simulation, and is immediately consumed to produce the species that are present at time 0.

Gillespie algorithm [5] and its various extensions provide an exact method for computing the reaction occurrences of CRNs. By using this algorithm, based on continuous time Markov chains, it is possible to run stochastic simulations. Such simulations can also be approximated by ordinary differential equations. However, stochastic simulations can give rise to observations that are otherwise impossible in a deterministic setting, as stochasticity provides a means for observing random fluctuations in species numbers. As an example, consider the CRN in Fig. 1, which is a Lotka-Volterra predator-prey system [12,18]. The algorithm for stochastic flux analysis builds on the Gillespie algorithm in a way that permits the tracking of individual species as they become consumed and produced throughout the simulation. By tracking these interactions, the

---

[1] https://sites.google.com/site/ozankahramanogullari/software.

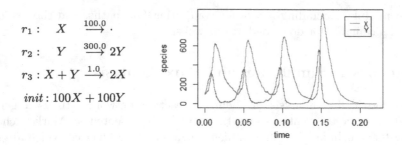

$$r_1 : \quad X \xrightarrow{100.0} .$$

$$r_2 : \quad Y \xrightarrow{300.0} 2Y$$

$$r_3 : X + Y \xrightarrow{1.0} 2X$$

$$init : 100X + 100Y$$

Fig. 1. A CRN model of a Lotka-Volterra predator-prey system. X represents a predator species, and Y represents a prey species. Unlike ordinary differential equation simulations, stochastic simulations as this one can capture spontaneous extinctions.

algorithm generates a quantitative log of dependencies between instances of reactions. The mechanism for this is realized by assigning a unique integer identifier to each individual species. The algorithm uses this information to incrementally construct an edge-colored graph structure by applying graph transformations. In this graph, the nodes are reactions of the CRN, and the edges are pairs of species and weights that quantify how many copies of which species flowed from which reaction to which other reaction, as exemplified in Fig. 2.

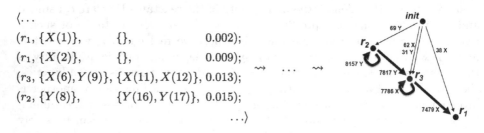

$$\langle \dots$$
$$(r_1, \{X(1)\}, \qquad \{\}, \qquad 0.002);$$
$$(r_1, \{X(2)\}, \qquad \{\}, \qquad 0.009);$$
$$(r_3, \{X(6), Y(9)\}, \{X(11), X(12)\}, 0.013); \quad \rightsquigarrow \quad \dots \quad \rightsquigarrow$$
$$(r_2, \{Y(8)\}, \qquad \{Y(16), Y(17)\}, 0.015);$$
$$\dots \rangle$$

Fig. 2. Besides the time series in Fig. 1, a simulation trajectory as the one on the left is generated from the CRN during simulation as described in the text. A number of graph transformations that are applied to this structure deliver the flux graph [10].

In [10], the construction of the flux graphs is realized in a number of steps. As the first step, each species instance is assigned a unique integer identifier in the initial state. Through out the simulation, each reaction instance randomly consumes the species that match its reactants, and are randomly selected from all the possibilities. The reaction instance then introduces its products to the current state with fresh integer identifiers. Each simulation step is recorded with respect to this information in the simulation log, as exemplified in Fig. 2 for the CRN in Fig. 1. By using the unique identifiers of the species in this structure, called *reaction trajectory*, the algorithm constructs a directed acyclic graph (dag) structure, where the nodes are species instances and the edges are the reaction instances that modify these edges.

By further processing this graph, the algorithm in [10] delivers an edge-labeled directed multi-graph that reveals the independence and causality information of the transitions with respect to the flow of specific resources between reactions. Since a reaction may produce several instances of species, this structure is in general a multi-graph. This dag, highlights the production-consumption relationship between reaction instances of the simulation, and this way provides a causality history of the simulation. Flux graphs, called *flux configurations*, are then obtained by compressing these dags in order to quantify the flow of resources between the reactions within given time intervals of the simulation. More specifically, the weight of each edge specifies the number of times the species on that edge flowed from the source to the target reaction of that edge.

In order to enable the recording of the simulation trajectory as described above, the reactions act on individual instances of species, rather than types of species as it is the case in the original Gillespie algorithm. Thus, a reaction of the CRN becomes a scheme, similar to a term rewriting rule. Although this modification does not introduce an increase in the complexity of the Gillespie algorithm, neither does it hamper its correctness with respect to the mass action kinetics, it introduces an overhead due to the individual representation of species. This overhead effects the simulation efficiency, and extends the simulation time in comparison to the standard Gillespie algorithm. In some cases with tens of thousands of individuals, it also introduces a limiting factor for running simulations due to the memory required to trace individuals.

In the following, we introduce an alternative algorithm that directly constructs the flux graphs during simulation by a minimal extension of the Gillespie algorithm, and this way avoids the overhead due to the labeling of the species.

3 Refining the Stochastic Flux Analysis Algorithm

Given a CRN, the Gillepsie algorithm [5], or the SSA, is a Monte Carlo simulation procedure that faithfully selects the next reaction j and its time τ. Thus, given a CRN, an initial state, and a t_{max}, by using this algorithm a time series s can be obtained for a time interval $0 \leq t \leq t_{max}$.

Let us consider a CRN with N species $\{S_1, ..., S_N\}$, which interact through M reactions $\{R_1, ..., R_M\}$. We denote with $\mathbf{X}(t) = (X_1(t), ..., X_N(t))$ the system state vector that represents the population of each S_i, whereby the CRN describes the time evolution of $X(t)$. The occurrence of each reaction R_j is then a discrete random event that changes the system state by $\mathbf{v}_j = \mathbf{p}_j - \mathbf{r}_j = (v_{1j}, ..., v_{Nj})$. The ith element v_{ij} specifies the change in X_i by one R_j reaction event, whereby p_{ij} specifies the products added to the state due to the right-hand-side of R_j, and r_{ij} specifies the reactants removed from the state due to the left-hand-side of R_j. Thus, given the system is in state $\mathbf{x} = (x_1, ..., x_N)$, the system jumps to state $\mathbf{x}' = \mathbf{x} + \mathbf{v}_j = \mathbf{x} + \mathbf{p}_j - \mathbf{r}_j$ as a consequence of a single R_j reaction event. The time that the next event of reaction R_j occurs is governed by function a_j, the propensity function of reaction R_j, with $a_0(\mathbf{x}) = \sum_{j=1}^{M} a_j(\mathbf{x})$, which are updated after each simulation step according to the new state $\mathbf{x}'$.

The refined algorithm, fSSA, for computing flux configurations of CRN simulations is a conservative extension of the SSA. The steps of the algorithm that extend SSA are denoted with '(·)' in Algorithm 1. The flux configuration is computed by updating two matrices at every simulation step. The algorithm initializes an $(M + 1) \times M \times N$ matrix $\mathbf{f}$ by setting all its cells to 0 (line 4). The matrix $\mathbf{f}$ delivers the simulation fluxes at the end of the simulation, as it is updated at every simulation step. The size $M + 1$ at the first dimension of $\mathbf{f}$ is due to M reactions and an additional reaction for the initial state; the size M at the second dimension is due to M reactions; the size N at the third dimension is due to N species. Then, each cell $\mathbf{f}_{\ell,j,i}$ denotes the number of species S_i that flow from R_ℓ to R_j, and the matrix $\mathbf{f}$ is output together with the time series s.

Algorithm 1. fSSA

Input: A CRN with N species and M reactions, initial state $\mathbf{x}_0$, and t_{max}.
Output: A time series s and a flux matrix $\mathbf{f}$.

1: $t \leftarrow 0$
2: $\mathbf{x} \leftarrow \mathbf{x}_0$
3: $s \leftarrow \langle \mathbf{x}_0 \rangle$
4: (·) $\mathbf{y} \leftarrow \mathbf{x}_0$
5: (·) initialize $\mathbf{f}$ such that all the cells are 0.
6: (·) initialize $\mathbf{m}$ such that $m_{0,i}$ is set as in x_i, and all others are 0.
7: evaluate all $a_j(\mathbf{x})$ and calculate $a_0(\mathbf{x})$
8:     **while** $t \leq t_{max}$ **do**
9:         $\tau \leftarrow$ a sample of exponential random variable with mean $1/a_0(\mathbf{x})$
10:        $u \leftarrow$ a sample of unit uniform random variable
11:        $\mu \leftarrow$ smallest integer satisfying $\Sigma_{i=1}^{\mu} a_i(\mathbf{x}) \geq u a_0(\mathbf{x})$
12:        $t \leftarrow t + \tau$
13:        $\mathbf{x} \leftarrow \mathbf{x} + v_\mu$
14:        $s \leftarrow s; \mathbf{x}$
15:        update $a_j(\mathbf{x})$, and recalculate $a_0(\mathbf{x})$
16: (·)   **for** $i = 1$ to N **do**
17: (·)   **for** $k = 1$ to $r_{i,\mu}$ **do**
18: (·) $w \leftarrow$ a sample of unit uniform random variable
19: (·) $\sigma \leftarrow$ smallest integer satisfying $\Sigma_{j=1}^{\sigma} \mathbf{m}_{i,j} \geq w y_i$
20: (·) $\mathbf{m}_{i,\sigma} \leftarrow \mathbf{m}_{i,\sigma} - 1$
21: (·) $y_i \leftarrow y_i - 1$
22: (·) $\mathbf{f}_{\sigma,\mu,i} \leftarrow \mathbf{f}_{\sigma,\mu,i} + 1$
23: (·)   **done**
24: (·) $\mathbf{m}_{i,\mu} \leftarrow \mathbf{m}_{i,\mu} + p_{i,\mu}$
25: (·) $y_i \leftarrow y_i + p_{i,\mu}$
26: (·)   **done**
27:    **end while**

The second matrix that the algorithm uses for book-keeping is an $N \times (M+1)$ matrix $\mathbf{m}$, which is initialized at the beginning of the simulation, and updated at every simulation step (line 5). In $\mathbf{m}$, there are $M + 1$ columns, because the first column denotes the initial state as a reaction. Thus, at time zero, the first column of $\mathbf{m}$ is initialized as the vector $\mathbf{x}_0$ and all the other cells are set to 0.

The fSSA algorithm is conservative of SSA, as it does not modify the SSA steps, and extends it with structures for logging the fluxes. The matrix $\mathbf{m}$ keeps track of the source reactions of species as they are being produced and consumed at every step. Each cell of the matrix displays a count of the species such that $\mathbf{m}_{ij}$ is the number of species S_i that had been produced by the reaction R_j, and had not been consumed by another reaction up to that point in the simulation. Thus, since each row carries the information on a certain species, each row sums up to the number of that species at the current state $\mathbf{x}$, that is, $\sum_{j=0}^{M} \mathbf{m}_{i,j} = x_i$. This information is used to sample the source of a species that can be produced by different reactions, proportional to the contribution of each reaction in producing that species (lines 15 to 22). The matrix $\mathbf{f}$ is updated accordingly (line 20). The products are then directly updated in $\mathbf{m}$ for the next simulation step (line 23).

The construction of the flux graph in Algorithm 1 introduces only a constant cost by introducing data structures that are accessed only for book-keeping, thus it is not subject to the overhead in the algorithm [10]. This is because, Algorithm 1 avoids labeling of the individual species, and this way permits the reactions to be applied on the multiplicities of instances instead of their actual occurrences.

4   Flux Paths

The FluxPath tool consists of two modules. The first module computes the flux configuration and saves this in a file. The flux configuration can be computed with the algorithm above or equivalently with the one in [10]. The former computes the flux configuration during simulation, whereas the latter first computes the simulation trajectory, which is saved to a separate file, and the flux configuration is then computed by processing this file. The second module takes the flux configuration as input and enumerates the pathways of information flow for various starting nodes and lengths of paths. The paths are computed by searching for paths in the flux configuration, which is an edge-colored weighted graph. The weight and the color, that is, the species, of each edge is kept as they are in the flux configuration during the search, and displayed in the output paths.

For an example consider the network depicted in Fig. 3 and its time series. The system implemented by this CRN is initiated in equilibrium. However, random fluctuations shift the system in a direction that favors either S2 or S5. In the simulation in Fig. 3, the dynamics results in large and small shifts at many occasions, which are visualized as fluctuations. After the time point around 550, the S2 production outweighs, which is observed as a rapid increase in S2 numbers.

By using our tool, we have analyzed the underlying dynamics with respect to the fluxes from the time point 550 to the end of the simulation; the flux graph is depicted in Fig. 4. In this graph, we observe that S4 and S6 fluxes between r_4 and r_5 have approximately the same weight, whereas S5 has a larger flux towards r_4 in comparison to S4 and S6 fluxes. Conversely, the fluxes between r_1 and r_2 weigh towards r_2. Moreover, a comparison of the S5 flux from r_3 to r_4

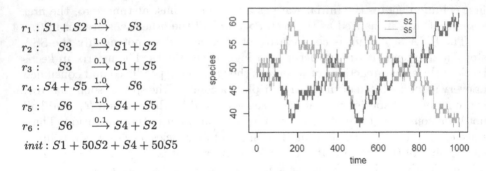

$r_1 : S1 + S2 \xrightarrow{1.0} S3$

$r_2 : \quad S3 \xrightarrow{1.0} S1 + S2$

$r_3 : \quad S3 \xrightarrow{0.1} S1 + S5$

$r_4 : S4 + S5 \xrightarrow{1.0} S6$

$r_5 : \quad S6 \xrightarrow{1.0} S4 + S5$

$r_6 : \quad S6 \xrightarrow{0.1} S4 + S2$

$init : S1 + 50S2 + S4 + 50S5$

Fig. 3. A CRN of two antagonist systems; S2 and S5 compete to break the equilibrium.

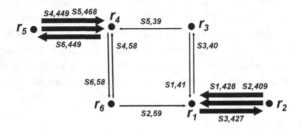

Fig. 4. The flux graph of the CRN in Fig. 3 for the time interval from 550 to 1000. The thickness of the arrows are proportional with strength of the fluxes.

and the S2 flux from r_6 to r_1 support a dynamics towards r_2. Finally, the higher turnover around r_6 in comparison to the turnover around r_3 supports the high r_2 activity that explains the shift of resources towards r_2, and the consequent excess in S2 observed in the time-series.

We have computed the paths of the flux configuration by using the FluxPath tool, available for download at our website. Among many other paths, the output flux paths depicted in Fig. 5 quantify the information flow between the S2 producing reaction r_2 and the S5 producing reaction r_5 after the time point 550.

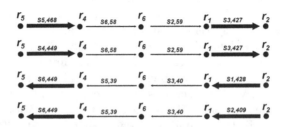

Fig. 5. The flux paths of the CRN in Fig. 3 between the S2 producing reaction r_2 and the S5 producing reaction r_5 for the time interval from 550 to 1000. The thickness of the arrows are proportional with the strength of the fluxes.

5   A Case Study: Gemcitabine

Gemcitabine (dFdC) is a prodrug, which is commonly used in the treatment of patients with non-small-cell lung cancer, pancreatic cancer, bladder cancer, and breast cancer. It is currently one of the leading therapeutic treatments for these diseases [3,4,17]. Gemcitabine exerts its clinical effect by depleting the deoxyribonucleotide pools, the building blocks of the DNA, and incorporating its triphosphate metabolite (dFdC-TP) into DNA, thereby inhibiting DNA synthesis. The incorporation of gemcitabine into DNA takes place in competition with the natural nucleotide dCTP, and this competition is an efficacy determining factor, which can be affected by various environmental and genetic conditions.

In [9], we have a given CRN model of gemcitabine biomolecular action, depicted in Fig. 6, which quantifies the the mechanisms of competition between the cascades that incorporate dCTP and dFdC-TP into the DNA. The simulations with this model identified certain mechanisms of crosstalk between these two pathways that affect the competition for incorporation. In agreement with the clinical studies dedicated to singling out mechanisms of resistance, our model associated ribonucleotide reductase (RR) and deoxycytidine kinase (dCK) activities to the efficacy of gemcitabine. Beside other mechanisms, such as transport across the plasma membrane, the inhibitory and enzymatic roles of these proteins determine efficacy depending on the availability of other metabolites.

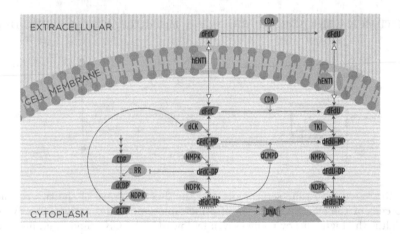

Fig. 6. The biochemical machinery of gemcitabine. Gemcitabine (dFdC and dFdU) is transported into cells by nucleoside transporters. It is then phosphorylated to its active diphosphate (dFdC-DP and dFdU-DP) and triphosphate (dFdC-TP and dFdU-TP) metabolites. Gemcitabine exerts its effect mainly by two mechanisms: while the diphosphate metabolite dFdC-DP plays an inhibitory role for the synthesis of natural nucleoside triphosphate dCTP, the triphosphate metabolite dFdC-TP competes with the dCTP for incorporation into nascent DNA chain, thereby inhibiting DNA synthesis and blocking cells in the early DNA synthesis phase. Image adopted from [9].

The efficiency of the inhibitions due to the association of dCTP with dCK and the association of dFdC-DP with RR play a key role in adjusting the efficacy. In this respect, simulations with our model have predicted a continuum of non-efficacy to high-efficacy regimes, where the levels of dFdC-TP and dCTP are coupled in a complementary manner. The complementary action, in which either dCTP or dFdC-TP make it to the DNA, is determined by the efficiency of the inhibitory associations of dCTP with dCK and dFdC-DP with RR. The extremes of this continuum are represented on one end, at the high efficacy regime, by low dCTP/dCK affinity and high dFdC-DP/RR affinity. On the other end, there is the low efficacy regime, given by high dCTP/dCK affinity and low dFdC-DP/RR affinity. Representative time series for these regimes are depicted in Figure 7.

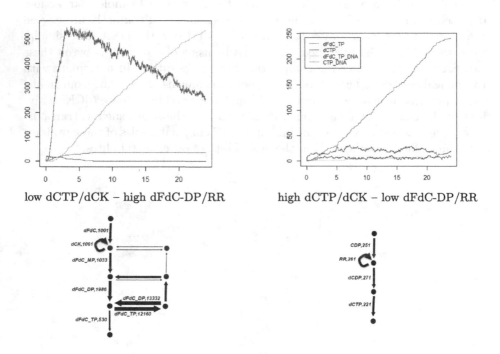

low dCTP/dCK – high dFdC-DP/RR   high dCTP/dCK – low dFdC-DP/RR

Fig. 7. Representative time series plots and the flux pathways of the two regimes at the two ends of the efficacy spectrum of gemcitabine molecular action. The dynamics on the left is the high efficacy regime given by low dCTP/dCK affinity and high dFdC-DP/RR affinity, whereas the one on the right is the low efficacy regime given by high dCTP/dCK affinity and low dFdC-DP/RR affinity.

We have performed flux analysis by using our tool on simulations in these regimes at either ends of the spectrum to quantify the effect of the inhibitory mechanisms on information flow from outside the cell into the DNA. Flux graphs of the dominant pathways for the two cases are depicted in Fig. 7.

In the low dCTP/dCK affinity and high dFdC-DP/RR affinity regime, the increase in association of dFdC-DP and RR depletes the RR pools, and as a result of this, RR becomes unavailable to serve as an enzyme for the cascade that incorporates dCTP into the DNA. Concomitantly, the decrease in association of dCTP and dCK increases the availability of dCK to serve as enzyme in the cascade that incorporates dFdC-TP into the DNA. This results in a dominant pathway of flux depicted on the left-hand-side of Fig. 7.

At the other end of the spectrum, in the high dCTP/dCK affinity and low dFdC-DP/RR affinity regime, the complementary mechanism depletes dCK pools due to increased association of dCTP and dCK. This hampers the pathway that would otherwise incorporate the dFdC-TP into the DNA. Moreover, as a consequence of the reduction in the association of dFdC-DP and RR, more RR becomes available to serve as enzyme in the pathway that incorporates dCTP into the DNA. The resulting dynamics delivers the pathway of flux depicted on the right-hand-side of Fig. 7.

6   Discussion

We have presented a method for flux analysis in stochastic simulations of chemical reaction networks that refines our previously published method [10]. In contrast to the method in [10], the algorithm here does not require the tracking of individual species, and monitors the network state during simulation in the form of species-type vectors as in the direct method [5]. The flux graphs are then computed by instantiating and updating two arrays, the size of which are bounded by the number of reactions and species-types of the CRN. Because the algorithm is not subject to an overhead due to the number of species, it can be applied to any CRN that can be simulated with the direct method, including those with arbitrarily small species numbers. As with time series plots of stochastic simulations, simulations with greater number of events provide more convergent observations, whereas smaller number of events highlight the stochastic nature of the systems due to random fluctuations. In this respect, the method for stochastic flux analysis provides a point of view for individual simulations that is complementary to their time series considerations. The computation of the flux graphs is not restricted to steady or stationary states, and it can be performed on arbitrary time intervals as demonstrated in our examples.

Our module for computing flux paths introduces a filter that is alternative to the global view of the flux graphs, as flux paths do not have any branching. In this respect, various filters such as cut-off thresholds or filtering out certain species in flux graphs can be considered for observations on different aspects of the CRNs. Other topics of future work include implementation of an integrated modeling suit that collects features above and others, as well as investigations with a more theoretical nature, in particular, the influence of different aspects of reaction networks such as the relative contribution of structure and non-linearity to the dynamical behavior of the system, and statistical queries that can provide insights to CRN dynamics.

Acknowledgments. This work has been partially funded by the European Union's Horizon 2020 research and innovation programme under the grant agreement No 686585 - LIAR, Living Architecture.

References

1. Cardelli, L., Kwiatkowska, M., Laurenti, L.: Stochastic analysis of chemical reaction networks using linear noise approximation. Biosystems **149**, 26–33 (2016)
2. Erhard, F., Friedel, C.C., Zimmer, R.: FERN - a java framework for stochastic simulation and evaluation of reaction networks. BMC Bioinform. **9**, 356 (2008)
3. Fryer, R.A., Barlett, B., Galustian, C., Dalgleish, A.G.: Mechanisms underlying gemcitabine resistance in pancreatic cancer and sensitisation by the iMiD^{TM} lenalidomide. Anticancer Res. **31**(11), 3747–3756 (2011)
4. Funel, N., Giovannetti, E., Chiaro, M.D., Mey, V., Pollina, L.E., Nannizzi, S., Boggi, U., Ricciardi, S., Tacca, M.D., Bevilacqua, G., Mosca, F., Danesi, R., Campani, D.: Laser microdissection and primary cell cultures improve pharmacogenetic analysis in pancreatic adenocarcinoma. Lab Invest. **88**(7), 773–784 (2008)
5. Gillespie, D.T.: Exact stochastic simulation of coupled chemical reactions. J. Phys. Chem. **81**(25), 2340–2361 (1977)
6. Gillespie, D.T.: A rigorous derivation of the chemical master equation. Physica A **188**, 404–425 (1992)
7. Gillespie, D.T.: Approximate accelerated stochastic simulation of chemically reacting systems. J. Chem. Phys. **115**(4), 1716 (2001)
8. Kahramanoğulları, O.: On linear logic planning and concurrency. Inf. Comput. **207**, 1229–1258 (2009)
9. Kahramanoğulları, O., Fantaccini, G., Lecca, P., Morpurgo, D., Priami, C.: Algorithmic modeling quantifies the complementary contribution of metabolic inhibitions to gemcitabine efficacy. PLoS ONE **7**(12), e50176 (2012)
10. Kahramanoğulları, O., Lynch, J.: Stochastic flux analysis of chemical reaction networks. BMC Syst. Biol. **7**, 133 (2013)
11. Kuwahara, H., Mura, I.: An efficient and exact stochastic simulation method to analyze rare events in biochemical systems. J. Chem. Phys. **129**(16), 10B619 (2008)
12. Lotka, A.J.: Fluctuations in the abundance of a species considered mathematically. Nature **119**, 12 (1927)
13. McQuarrie, D.A.: A rigorous derivation of the chemical master equation. J. Appl. Probab. **4**, 413–478 (1967)
14. Nielsen, M., Plotkin, G., Winskel, G.: Event structures and domains, part 1. Theor. Comput. Sci. **5**(3), 223–256 (1981)
15. Okino, M.S., Mavrovouniotis, M.L.: Simplification of mathematical models of chemical reaction systems. Chem. Rev. **98**(2), 391–408 (1998)
16. Salis, H., Kaznessis, Y.N.: Accurate hybrid stochastic simulation of a system of coupled chemical or biochemical reactions. J. Chem. Phys. **122**(5), 54103 (2005)
17. Veltkamp, S.A., Beijnen, J.H., Schellens, J.H.: Prolonged versus standard gemcitabine infusion: translation of molecular pharmacology to new treatment strategy. Oncologist **13**(3), 261–276 (2008)
18. Volterra, V.: Fluctuations in the abundance of species considered mathematically. Nature **118**, 558–560 (1926)

Parallel Biological Sequence Comparison in Linear Space with Multiple Adjustable Bands

Gabriel H.G. Silva, Edans F.O. Sandes, George Teodoro,
and Alba C.M.A. Melo$^{(\boxtimes)}$

Department of Computer Science, University of Brasilia (UnB), Brasilia, Brazil
alves@unb.br

Abstract. In this paper, we propose and evaluate Fickett-MM, a parallel strategy that combines the algorithms Fickett and Myers-Miller, splitting a pairwise sequence comparison into multiple comparisons of subsequences and calculating an appropriate Fickett band to each subsequence comparison (block). With this approach, we potentially reduce the number of cells calculated in the dynamic programming matrix when compared to Fickett, which uses a unique band to the whole comparison. Our adjustable multi-block strategy was integrated to the stage 4 of CUDAlign, a state-of-the-art parallel tool for optimal biological sequence comparison. Fickett-MM was used to compare real DNA sequences whose sizes ranged from 10KBP (Thousands of Base Pairs) to 47MBP (Millions of Base Pairs), reaching a speedup of 59.60× in the 10MBP × 10MBP comparison when compared to CUDAlign stage 4.

Keywords: Parallel biological sequence comparison · Multiple adjustable bands

1 Introduction

Pairwise biological sequence comparison is a widely used operation in Bioinformatics. It produces as output a score, which represents the similarity between the sequences, and an alignment [1]. The optimal global alignment with linear gap can be obtained with the Needleman-Wunsh (NW) algorithm [9], which is based on dynamic programming (DP) and has $O(mn)$ time and space complexity, where m and n are the lengths of the sequences. Smith-Waterman (SW) [14] proposed a DP-based algorithm that computes optimal local alignments with linear gap with the same time and space complexity.

Gotoh [3] modified the NW algorithm, calculating optimal alignments with the affine gap model. Since gaps tend to occur together in nature, the affine gap model is more appropriate for realistic scenarios. Hirschberg [4] proposed a variant of the NW algorithm that retrieves optimal alignments in linear space ($O(m + n)$) with linear gap. This variant was further modified by Myers-Miller (MM) [8] in order to use the affine gap model. Fickett [2] proposed an algorithm that retrieves the optimal global alignment by calculating only a k-band of the

© Springer International Publishing AG 2017
D. Figueiredo et al. (Eds.): AlCoB 2017, LNBI 10252, pp. 167–179, 2017.
DOI: 10.1007/978-3-319-58163-7_12

DP matrix near the main diagonal, where k is the number of diagonals computed, executing thus in $O(kn)$ time and space. If the alignment does not fall into the k-band, the band is enlarged and the DP matrix is iteratively re-computed until the alignment can be retrieved.

The use of the NW and SW algorithms and its variants to compare long DNA sequences or a protein sequence to a huge genomic database can lead to very high execution times and, for this reason, parallel solutions are usually employed. In the literature, there are many proposals that execute NW or SW variants in parallel architectures such as clusters [7,12], FPGAs (Field Programmable Gate Arrays) [13,16], GPUs (Graphics Processing Units) [6,10] and Intel Xeon Phis [5,15], among others. CUDAlign 4.0 [10] is a state-of-the-art tool which computes optimal local alignments between long DNA sequences in linear space using 5 stages. Stage 1 executes phase 1 of the Gotoh algorithm (score calculation) with affine-gap and stages 2 to 5 execute phase 2 (traceback), with an adapted version of the MM algorithm.

In this paper, we propose and evaluate Fickett-MM, a parallel strategy which combines MM with a variant of Fickett's [2]. Unlike the original Fickett algorithm, we divided the alignment problem into several parts and computed a different k-band for each part of the problem, which is adjusted to its alignment characteristics. Fickett-MM was implemented in C++/pthreads and integrated to the stage 4 of CUDAlign. The results obtained with real DNA sequences whose sizes varied from 10 KBP to 47 MBP show that our strategy is able to achieve a speedup of up to 59.60× in stage 4 of CUDAlign, when compared to the original implementation. In the longest comparison, the execution time of CUDAlign stage 4 was reduced from 2 min and 54 s to 30 s.

The remainder of this paper is organized as follows. Section 2 presents algorithms for optimal biological sequence alignment and the CUDAlign tool is presented in Sect. 3. The design of Fickett-MM is explained in Sect. 4. In Sect. 5, experimental results are discussed and Sect. 6 concludes the paper.

2   Biological Sequence Comparison

2.1   Basic Algorithms - NW and SW

The Needleman-Wunsh (NW) [9] algorithm is based on DP and retrieves the optimal global alignment in $O(mn)$ space and time, executing in two phases: (a) calculate the DP matrix and (b) retrieve the alignment (traceback).

The first phase receives sequences S_0 and S_1, with lengths n and m, and computes the DP matrix H as follows. The first row and column are filled with $H_{i,0} = i * g$ and $H_{0,j} = j * g$, where g is the gap penalty and i and j represent the sizes of the prefixes of the sequences. The remaining cells are calculated with the recurrence relation expressed by Eq. (1) [9].

$$H_{i,j} = \max \begin{cases} H_{i-1,j-1} + p(i,j) \\ H_{i,j-1} - g \\ H_{i-1,j} - g \end{cases} \tag{1}$$

	*	T	A	G	T	C			*	T	A	G	T	C
*	0	-2	-4	-6	-8	-10		*	0	0	0	0	0	0
T	-2	1	-1	-3	-5	-7		T	0	1	0	0	1	0
A	-4	-1	2	0	-2	-4		A	0	0	2	0	0	0
G	-6	-3	0	3	1	-1		G	0	0	0	3	1	0
C	-8	-5	-2	1	2	2		C	0	0	0	1	2	2

T	A	G	T	C		T	A	G
T	A	G	-	C		T	A	G
(a)						(b)		

Fig. 1. DP matrices and alignments for S_0 and S_1 (mi $= -1$, ma $= +1$, g $= -2$). (a) NW matrix; (b) SW matrix.

In this equation, if DNA or RNA sequences are compared, $p(i,j)$ is the match punctuation (ma), if $S_0[i] = S_1[j]$, or the mismatch punctuation (mi), otherwise. If amino acid sequences are compared, $p(i,j)$ is given by a given 20×20 substitution matrix [1]. Each cell $H_{i,j}$ keeps an indication of which cell $(H_{i-1,j-1})$, $(H_{i,j-1})$ or $(H_{i-1,j})$ was used to produce its value (arrows in Fig. 1). The optimal score is in cell $H_{m,n}$. In order to produce the alignment, phase 2 (traceback) is executed from the bottom right cell in the DP matrix, following the indications until the top left cell is attained. Figure 1(a) illustrates the DP matrix calculated by NW. The optimal score is 2 and the optimal global alignment, obtained in the traceback phase, is shown below the DP matrix.

When the biologists are interested in calculating how similar the fragments of the sequences are, local alignment is usually applied and the Smith-Waterman (SW) algorithm is used. The SW uses DP, has the same complexity of NW and executes in two phases. Nevertheless, NW and SW are distinct in three ways. First, differently from NW, the elements of the first row and column of the SW matrix are set to zero. Second, the SW recurrence relation is slightly different from the NW recurrence relation since no negative values are allowed in SW (Eq. (2)) [14]. Finally, the cell that contains the optimal local score is the cell $H_{i,j}$ which has the highest value in H. In the traceback phase, SW starts from cell $H_{i,j}$, following the arrows until a cell whose value is zero is found. Figure 1(b) illustrates the DP matrix calculated by SW. In this figure, the optimal score is 3 and the optimal local alignment is shown below the DP matrix.

$$H_{i,j} = \max \begin{cases} H_{i-1,j-1} + p(i,j) \\ H_{i,j-1} - g \\ H_{i-1,j} - g \\ 0 \end{cases} \tag{2}$$

2.2   NW and SW Variants

To produce more biologically relevant results, Gotoh [3] proposed an algorithm that implements the affine-gap model, with two different gap penalties:

one to initiate a sequence of gaps (G_{first}) and another to extend it (G_{ext}). Gotoh calculates three DP matrices: H, E and F, where H keeps track of matches/mismatches and E and F keep track of gaps in each sequence (Eqs. 3, 4 and 5).

$$H_{i,j} = \max \begin{cases} 0 \\ E_{i,j} \\ F_{i,j} \\ H_{i-1,j-1} - p(i,j) \end{cases} \tag{3}$$

$$E_{i,j} = \max \begin{cases} E_{i,j-1} - G_{ext} \\ H_{i,j-1} - G_{first} \end{cases} \tag{4}$$

$$F_{i,j} = \max \begin{cases} F_{i-1,j} - G_{ext} \\ H_{i-1,j} - G_{first} \end{cases} \tag{5}$$

When long sequences are compared, linear space algorithms should be used. One of the first linear space algorithms for sequence comparison is the one proposed by Hirschberg [4]. First, the DP matrix is calculated from the beginning to the middle row ($i*$), storing only the last row calculated. After that, the DP matrix is calculated from the end to $i*$, over the reverses of the sequences. At this point, there are two middle rows, one calculated with the original sequences and another one calculated with the reverses of the sequences. Hirschberg proved that the position where the addition of the corresponding values in these two middle rows is maximum belongs to the optimal alignment [4]. This point is called crosspoint and it divides the problem into two smaller subproblems, which are processed recursively, until trivial solutions are found. Myers-Miller (MM) [8] adapted Hirschberg to the Gotoh algorithm by using two additional vectors. The first and second recursions of the MM algorithm are shown in Fig. 2.

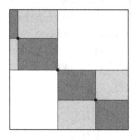

Fig. 2. Myers-Miller (MM) algorithm. The black circles represent the crosspoints.

Fickett [2] proposed an algorithm that can be executed quickly if the sequences compared are very similar. In this case, the alignment between the sequences is confined in a small region near the main diagonal of the DP matrix.

Thus, Fickett only calculates and stores a small set of diagonals near the main diagonal (k-band), with time and space complexity $O(kn)$. The k-band is estimated with a heuristic measurement of the similarity of the sequences. The optimal score is contained in cell $H_{n,m}$ and it is used to do the traceback over the band. If the k-band was underestimated, the alignment cannot be retrieved (Fig. 3a). In this case, the algorithm enlarges iteratively the k-band and the DP matrix is calculated for the new k-band, until the whole alignment is obtained (Fig. 3b).

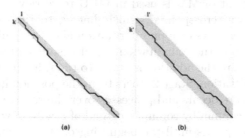

Fig. 3. Fickett's algorithm. The gray area represents the k-band.

Although algorithms MM and Fickett have been proposed to the global alignment problem, they can be easily adapted to the local alignment case as follows. First, the DP matrix is processed with SW, giving as output the highest score. Second, the matrix is recalculated from the position where the optimal score occurs over the reverses of the sequences until the position where the optimal local alignment begins is found. With these two positions, the problem is transformed into a global alignment problem and MM or Fickett can be readily applied.

3   Design of CUDAlign 4.0

CUDAlign [10] is a tool that obtains the optimal local alignment between two long DNA sequences in GPU, using adapted versions of the Gotoh and MM algorithms (Sect. 2.2). CUDAlign executes in 5 stages, as shown in Fig. 4.

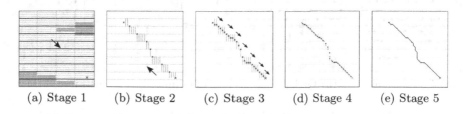

(a) Stage 1   (b) Stage 2   (c) Stage 3   (d) Stage 4   (e) Stage 5

Fig. 4. General overview of CUDAlign

Stage 1 corresponds to the first phase of the Gotoh algorithm and executes in GPU in linear space, giving as output the highest score and its position in the matrix. Stage 1 uses mainly two optimizations. The optimization *cells delegation* processes the Gotoh matrices in multiple blocks, in a parallelogram wavefront shape, allowing maximum parallelism most of the time. The optimization *block pruning*, shown in gray in Fig. 4(a), does not compute DP cells which certainly do not contribute to the optimal alignment. In order to accelerate the further stages, some rows of the DP matrices (special rows) are stored. Stages 2, 3, 4 and 5 implement phase 2 (traceback).

In stage 2, a variant of MM is used in GPU to retrieve the midpoints that cross the special rows (crosspoints), which belong to the optimal alignment. Unlike MM, the special rows contain information about the maximum score and can be used to accelerate the computation. So, it is sufficient to find the position in the special row where the addition is equal to the (already known) maximum score. With this observation, Stage 2 starts from the position in the DP matrix in which the optimal score occurs and processes over the reverses of the sequences, calculating the area column by column (instead of row by row, as in the original MM) and finding midpoints until the beginning of the optimal local alignment is found. In stage 3, the beginning and end of the optimal local alignment are received as input. Moreover, the special columns saved to disk in stage 2 are used. Stage 3 starts from the beginning of the alignment and uses the special columns to retrieve more crosspoints in GPU.

Stage 4 executes in CPU using a modified MM algorithm between each successive pair of crosspoints (partition) found on stage 3, with multiple threads. The goal of stage 4 is to increase the number of crosspoints until the distance between any successive pair of crosspoints is smaller than a given limit (e.g. 16×16). Figure 4(d) shows the additional crosspoints obtained in stage 4.

Figure 5 presents four ways to process a partition in stage 4, where the gray areas represent the regions that do not need to be processed after the crosspoint is found. The conventional MM algorithm processes both halves of the partition entirely (Fig. 5(a)). CUDAlign 2.0 introduced an optimization called Orthogonal Execution, which processes the top half of the partition over rows and the bottom half of the partition over columns (Fig. 5(b)). CUDAlign 4.0 extended this idea processing both halves of the partition over columns, alternating columns from

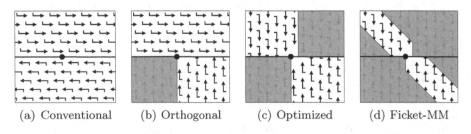

(a) Conventional   (b) Orthogonal   (c) Optimized   (d) Ficket-MM

Fig. 5. DP submatrix computed in stage 4. Area in gray are not processed.

each half (Fig. 5(c)). The Ficket-MM algorithm proposed in this paper (Sect. 4) reduces further the area processed (Fig. 5(d)).

Stage 5 aligns in CPU each partition found in stage 4 using NW. Then it concatenates all the results, giving as output the full optimal alignment (Fig. 4(e)). Stage 6 is an optional stage used only for visualization of the alignment.

4   Design of Fickett-MM

The main goal of Fickett-MM is to reduce the area computed in the alignment retrieval. In order to achieve this goal, we combine the well-known algorithms Fickett and MM (Sect. 2.2), creating the notion of adjustable bands. We assume that, as in MM, the computation is divided into blocks and the scores at the top left and bottom right corners of a block are known. With this information, we are able to define computation bands of different sizes, one for each block, in which the optimal alignment is guaranteed to be found.

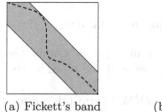   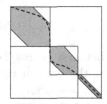

(a) Fickett's band   (b) Fickett-MM adjustable bands

Fig. 6. Bands in the Fickett algorithm and in the Fickett-MM algorithm

In Fig. 6, we illustrate the main difference between Fickett-MM and the original Fickett algorithm. In Fig. 6(a), Fickett's band (gray area) must encompass the whole alignment (dashed line), which has a considerable number of gaps in its beginning. For this reason, the size of the band is big, even though the alignment does not have many gaps in its end. On the other hand, Fickett-MM (Fig. 6(b)) defines three different bands (gray area), one for each block.

The efficiency of Fickett-MM is highly dependent on a good estimation of size of the computation bands. The scores at the upper left corner ($score_l$) and at the bottom-right corner ($score_r$) are known and a block is the rectangle defined by these two points.

The size of the band for each block is computed with Eq. 6 and it depends on four terms: PM, $score_d$, DPM and min_g, which are explained in the following paragraphs.

$$band = \left\lceil \frac{PM - score_d}{DPM} \right\rceil + min_g \qquad (6)$$

The perfect match term (PM) computes the maximum score of the block in the best case, i.e., all the corresponding characters of the subsequences are

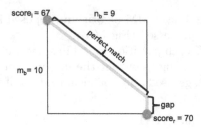

Fig. 7. Elements used in the perfect match (PM) computation.

the same (perfect match). Since the lengths of the subsequences may not be the same (Fig. 7), the length of the smaller subsequence is multiplied by the match punctuation and subtracted by the difference on the lengths of the subsequences multiplied by the gap penalty. In Eq. 7, ma is the punctuation for matches and gap is the punctuation for gap extension.

$$PM = min(m_b, n_b) * ma - (max(m_b, n_b) - min(m_b, n_b)) * |gap| \qquad (7)$$

The difference between scores ($score_d$) in Eq. 6 is simply the difference between the score at the bottom right corner and the score at the upper left corner: $score_d = score_r - score_l$.

The deviation from the perfect match (DPM) term takes into account the fact that each time a gap is introduced, we need at least another gap to return to the perfect match case, and, since two gaps are introduced, one punctuation for match (ma) will not be counted. Equation 8 presents this computation.

$$DPM = |2 * gap| + ma \qquad (8)$$

Finally, the term min_g calculates the difference between the sizes of the subsequences (m_b and n_b) since it indicates the minimum number of gaps needed for the band to contain the optimal alignment (Eq. 9).

$$min_g = max(m_b, n_b) - min(m_b, n_b) \qquad (9)$$

The size of the band is computed by considering the worst case, i.e., gaps are introduced instead of mismatches. In addition, since we do not know in which sequence gaps will be introduced, we apply the same value of band for both sides of the perfect match case. With this, we guarantee that the band encompasses the optimal alignment even though in some cases it will be larger than necessary. In order to illustrate the computation of the size of the band, consider the values in Fig. 7 and assume that $ma = +1$, $gap = -2$. In this case, $PM = 7$, $score_d = 3$, $DPM = 5$ and $min_g = 1$, giving 2 as the size of the band in each side of the perfect match case (Eq. 6).

Algorithm 1 presents the pseudocode of Fickett-MM. It receives as input the subsequences S_0' and S_1' as well as the scores in the upper left and bottom right

---

Algorithm 1. Fickett-MM

Require: Subsequences S_0' e S_1', Scores $score_l$ and $score_r$
Ensure: $crosspoint$
 1: /*Calculates the size of the band*/
 2: $score_d \leftarrow score_r - score_l$
 3: $k \leftarrow calculate\_band(S_0', S_1', score_d)$
 4: $seq1\_length \leftarrow size(S_0')$
 5: $j \leftarrow 0$
 6: **loop**
 7:      /*Calculates the extremities of the columns inside the band*/
 8:      $upper\_left \leftarrow Calculate\_FickettMM\_upper\_left(j, k)$
 9:      $lower\_left \leftarrow Calculate\_FickettMM\_lower\_left(j, k)$
10:      $upper\_right \leftarrow Calculate\_FickettMM\_upper\_right(seq1\_length - j, k)$
11:      $lower\_right \leftarrow Calculate\_FickettMM\_lower\_right(seq1\_length - j, k)$
12:      /*Calculates the recurrence equation inside the band*/
13:      $crosspoint1[j] \leftarrow Compute\_FickettMM(upper\_left, lower\_left, S_0', S_1')$
14:      $crosspoint2[j] \leftarrow Compute\_FickettMM(upper\_right, lower\_right, S_0', S_1')$
15:      **if** $j > seq1\_length/2$ **then**
16:          $crosspoint \leftarrow crosspoint1[j] + crosspoint2[seq1\_length - j]$
17:          **if** $check(crosspoint, score_d) = TRUE$ **then**
18:              **return** $crosspoint$
19:          **end if**
20:          $crosspoint \leftarrow crosspoint2[j] + crosspoint1[seq1\_length - j]$
21:          **if** $check(crosspoint, score_d) = TRUE$ **then**
22:              **return** $crosspoint$
23:          **end if**
24:      **end if**
25:      $j + +$
26: **end loop**

---

cells ($score_l$ and $score_r$). The computation of the size of the band is done in lines 2 and 3 and its value is stored in k. Then, a loop is executed from lines 6 to 26 for every column j as follows. In lines 8 to 11, the algorithm calculates the extremeties of column j (forward direction) and column $seq1\_length - j$ (reverse direction) up to the middle row. Then, the recurrence equation is calculated for both columns j and $seq1\_length - j$, as illustrated in Fig. 5(d). The values of the cells in the middle row are stored in vectors $crosspoint1$ (line 13) and $crosspoint2$ (line 14). When the middle column is attained (line 15), crosspoints 1 and 2 are added accordingly (lines 16 and 20) and the algorithm checks if the results match $score_d$ (i.e. $score_r - score_l$). If one of these values match (lines 17 and 21), this crosspoint is returned (lines 20 and 24).

5 Experimental Results

Fickett-MM was implemented in C/C++/pthreads and integrated to the stage 4 of CUDAlign 4.0. In our tests, we used a desktop with a CPU Intel Core i7 3770 (4 hardware cores), 8 GB RAM, 1 TB disk and a GPU NVidia GTX 680 (1536 cores and 2 GB RAM).

The following parameters were used in the tests: *ma (match)* = +1, *mi (mismatch)* = −3, G_{first} *(First gap)* = −5, G_{ext} *(Gap extension)* = −2, *number of threads* = 8 and *final size of block* = 24 × 24.

The experiments used real DNA sequences, retrieved from the NCBI (National Center for Biotechnology Information) at www.ncbi.nlm.nih.gov.

Table 1. Sequences used in the tests.

Comparison size	Sequence S_0		Sequence S_1	
	Accession	Name	Accession	Name
10K × 10K	AF133821.1	HIV-1 isolate MB2059	AY352275.1	HIV-1 isolate SF33
57K × 57K	AF494279.1	C. globosum	NC_001715.1	A. macrogynus
162K × 172K	NC_000898.1	H. herpesvirus 6B	NC_007605.1	H. herpesvirus 4
543K × 536K	NC_003064.2	A. fabrum C58	NC_000914.1	Rhizobium sp. NGR234
1M × 1M	CP000051.1	C. trachomatis	AE002160.2	C. muridarum
3M × 3M	BA000035.2	C. efficiens	BX927147.1	C. glutamicum
5M × 5M	AE016879.1	B. anthracis str. Ames	AE017225.1	B. anthracis str. Sterne
7M × 5M	NC_005027.1	R. baltica SH	AE016879.1	B. anthracis str. Ames
10M × 10M	NC_017186.1	A. mediterranei S699	NC_014318.1	A. mediterranei U32
23M × 25M	NT_033779.4	D. melanogaster chr. 2L	NT_037436.3	D. melanogaster chr. 3L
47M × 32M	NC_000021.7	H. sapiens chr. 21	BA000046.3	P. troglodytes chr. 22

Table 2. Execution time, speedup and characteristics of the alignment

Comparison	Fickett-MM (ms)	CUDAlign stage 4 (ms)	Speedup	Local score	Matches %	Mismatches %	Gaps %
10K × 10K	98.08	179.62	1.83×	5,091	89.12	9.64	1.24
57K × 57K	0.76	0.80	1.03×	80	92.50	5.00	2.50
162K × 172K	0.83	0.82	0.99×	18	100.00	0.00	0.00
543K × 536K	1.96	2.07	1.06×	48	88.04	11.96	0.00
1M × 1M	1,403.09	2,555.49	1.81×	88,535	79.76	17.12	3.12
3M × 3M	109.69	146.61	1.34×	4,226	83.05	10.46	6.49
5M × 5M	510.59	26,892.05	52.67×	5,220,960	99.95	0.00	0.05
7M × 5M	3.49	4.84	1.39×	172	84.07	12.74	3.19
10M × 10M	898.65	53,563.24	59.60×	10,235,188	99.99	0.01	0.00
23M × 25M	10.15	182.03	17.93×	9,063	99.88	0.05	0.07
47M × 32M	30,425.82	174,147.98	5.72×	27,206,434	94.38	1.54	4.08

Table 1 shows the accession number, the name and approximate size of each sequence.

Table 2 shows the execution times and speedups comparing Fickett-MM with CUDAlign stage 4 (optimized version). It can be seen in this table that, as expected, the best speedups are obtained when the sequences have high similarity, i.e., the local score is close to the size of the smallest sequence (Table 1). For the 5M × 5M and 10M × 10M comparisons, Fickett-MM executed more than 50 times faster than CUDAlign stage 4.

The comparison 23M × 25M obtained a high speedup (17.93×) even though the alignment is not so big. This suggests that, besides the similarity between the sequences, the shape of the alignment has a high influence over the speedups, as shown in the columns 5 to 8 in Table 2. It can be seen that alignments which have a high percentage of matches (>99%) have impressive speedups, with the exception of very small alignments (e.g. 162k × 172k comparison).

Table 3. Number of blocks vs. size of the band

Comparison	Band size (%)										Number of blocks
	0–10	10–20	20–30	30–40	40–50	50–60	60–70	70–80	80–90	90–100	
10k × 10k	108	**241**	116	70	18	10	4	4	2	1	574
57k × 57k	0	**2**	1	0	0	0	0	0	0	0	3
162k × 172k	0	0	0	0	0	0	0	0	0	0	0
543k × 536k	0	1	**2**	0	0	0	0	0	0	0	3
1M × 1M	1056	3538	6469	**9940**	3615	2677	841	357	241	913	29647
3M × 3M	**439**	179	74	69	60	59	38	58	65	164	1205
5M × 5M	**323193**	119	28	25	17	8	8	9	9	52	323468
7M × 5M	0	5	**14**	4	4	2	1	1	0	0	33
10M × 10M	**642225**	76	6	7	3	6	1	0	3	0	642327
23M × 25M	**507**	3	1	0	0	0	0	0	0	0	511
47M × 32M	**1788870**	151556	25926	16259	6943	5670	4212	2724	2734	25079	2029973

A detailed analysis of the alignment's shapes is presented in Table 3 and Fig. 8, showing the number of blocks that were processed with a given band size. The band sizes are given in percentage, calculated as the absolute size of the band divided by the size of the subsequence. For instance, if the size of the subsequence is 100 nucleotides and the size of the band is 12, the percentage is 12% and this block is counted in column "10–20%".

It can be seen that the comparisons in which Fickett-MM achieved impressive speedups (5M × 5M, 10M × 10M and 23M × 25M) only processed less than 10% of their blocks. The comparison 47M × 32M achieved a very good speedup (5.72×) but not as impressive as the three comparisons previously cited because of some blocks in which the size of the band is big. The 543K × 536K comparison

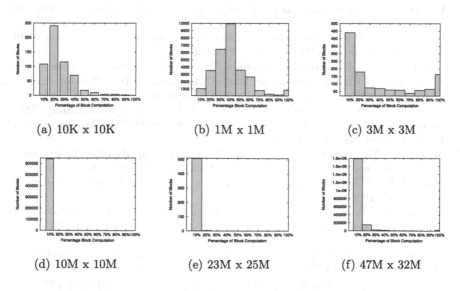

(a) 10K x 10K   (b) 1M x 1M   (c) 3M x 3M

(d) 10M x 10M   (e) 23M x 25M   (f) 47M x 32M

Fig. 8. Percentage of block computation for 6 comparisons

is a very interesting case in which the alignment is so small (size $= 18$) that it does not fill one entire block.

6   Conclusion

In this paper, we proposed and evaluated Fickett-MM, a strategy that retrieves the optimal alignment between two biological sequences using multiple adjustable Fickett bands in linear space. In order to compute the size of each band, we proposed a formula that uses the best score computed so far in special rows/columns, guaranteeing that the optimal alignment will be encompassed by the band. The computation of the adjustable bands was integrated to CUDAlign stage 4, a modified and parallel version of Myers-Miller, which retrieves optimal alignments in linear space.

The results obtained with sequence comparisons whose sizes ranged from $10K \times 10K$ to $47M \times 32M$ show that Fickett-MM is able to attain impressive speedups when the alignment is huge and the sequences are very similar. In the $10M \times 10M$ comparison, the execution time was reduced from $53.5\,s$ to $0.89\,s$.

As future work, we intend to port Fickett-MM to GPUs (CUDA and OpenCL). Also, we intend to adapt Fickett-MM to retrieve global and semi-global alignments, integrating it to the MASA tool [11].

References

1. Durbin, R., Eddy, S.R., Krogh, A., Mitchison, G.: Biological Sequence Analysis: Probabilistic Models of Proteins and Nucleic Acids. Cambridge University Press, Cambridge (1999)
2. Fickett, J.W.: Fast optimal alignments. Nucleic Acids Res. **11**, 175–179 (1984)
3. Gotoh, O.: An improved algorithm for matching biological sequences. J. Mol. Biol. **162**(3), 705–708 (1982)
4. Hirschberg, D.S.: A linear space algorithm for computing maximal common subsequences. Commun. ACM **18**(6), 341–343 (1975)
5. Liu, Y., Tam, T., Lauenroth, F., Schmidt, B.: SWAPHI-LS: Smith-Waterman algorithm on Xeon Phi coprocessors for long DNA sequences. In: IEEE International Conference on Cluster Computing, pp. 257–265 (2014)
6. Liu, Y., Wirawan, A., Schmidt, B.: CUDASW++ 3.0: accelerating Smith-Waterman protein database search by coupling CPU and GPU SIMD instructions. BMC Bioinformatics **14**, 117 (2013)
7. Maleki, S., Musuvathi, M., Mytcowicz, T.: Parallelizing dynamic programming through rank convergence. In: 19th ACM PPoPP, pp. 219–232 (2014)
8. Myers, E.W., Miller, W.: Optimal alignments in linear space. Comput. Appl. Biosci. **4**(1), 11–17 (1988)
9. Needleman, S.B., Wunsch, C.D.: A general method applicable to the search for similarities in the amino acid sequence of two proteins. J. Mol. Biol. **48**(3), 443–453 (1970)
10. de Oliveira Sandes, E.F., Miranda, G., Martorell, X., Ayguade, E., Teodoro, G., de Melo, A.C.M.: CUDAlign 4.0: incremental speculative traceback for exact chromosome-wide alignment in GPU clusters. IEEE Tran. Parallel Dist. Syst. **27**(10), 2838–2850 (2016)

11. de Oliveira Sandes, E.F., Miranda, G., Martorell, X., Ayguade, E., Teodoro, G., de Melo, A.C.M.: MASA: a multiplatform architecture for sequence aligners with block pruning. ACM Trans. Parallel Comput. **2**(4), 28 (2016)
12. Rajko, S., Aluru, S.: Space and time optimal parallel sequence alignments. IEEE Trans. Parallel Distrib. Syst. **15**(12), 1070–1081 (2004)
13. Sarkar, S., Kulkarni, G.R., Pande, P.P., Kalyanaraman, A.: Network-on-chip hardware accelerators for biological sequence alignment. IEEE Trans. Comput. **59**(1), 29–41 (2010)
14. Smith, T.F., Waterman, M.S.: Identification of common molecular subsequences. J. Mol. Biol. **147**(1), 195–197 (1981)
15. Wang, L., Chan, Y., Duan, X., Lan, H., Meng, X., Liu, W.: XSW: accelerating biological database search on Xeon Phi. In: IEEE AsHES, pp. 950–957 (2014)
16. Wienbrandt, L.: The FPGA-based high-performance computer RIVYERA for applications in bioinformatics. In: Beckmann, A., Csuhaj-Varjú, E., Meer, K. (eds.) CiE 2014. LNCS, vol. 8493, pp. 383–392. Springer, Cham (2014). doi:10.1007/978-3-319-08019-2_40

Author Index

Printed in the United States
By Bookmasters